Scotland
DEFENDING THE NATION

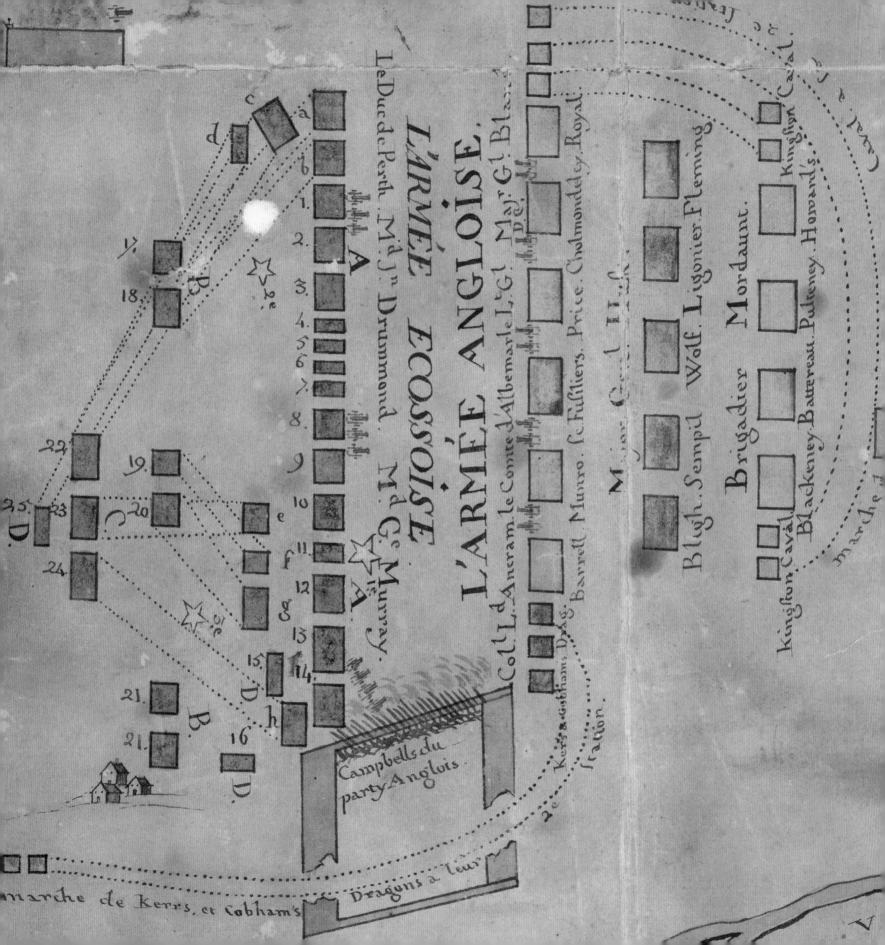

Scotland
DEFENDING THE NATION
MAPPING THE MILITARY LANDSCAPE

Carolyn Anderson and Christopher Fleet

BIRLINN

in association with the National Library of Scotland

First published in
Great Britain in 2018 by
Birlinn Ltd
West Newington House
10 Newington Road
Edinburgh
EH9 1QS

in association with the
National Library of Scotland

www.birlinn.co.uk

ISBN: 978 1 78027 493 5

Copyright © Carolyn Anderson
and Christopher Fleet 2018

The right of Carolyn Anderson and
Christopher Fleet to be identified
as the authors of this work has been
asserted by them in accordance
with the Copyright, Designs and
Patents Act, 1988

All rights reserved. No part of this
publication may be reproduced,
stored, or transmitted in any
form, or by any means, electronic,
mechanical or photocopying,
recording or otherwise, without
the express written permission
of the publisher.

*British Library
Cataloguing-in-Publication Data*
A catalogue record for this book
is available on request from the
British Library

Typeset and designed by
Mark Blackadder

PREVIOUS PAGE

*Anon., Plan exact de la disposition
des Troupes Ecossoises sous le
Commandement de son A. R. P. C.
et de Celle des Troupes Angloises
a la Bataille de Culloden prés la
Ville d'Inverness le 16e d'avril
1746 (1748).*

Printed and bound by PNB, Latvia

CONTENTS

	Preface and acknowledgements	vi
CHAPTER 1	Introduction	1
CHAPTER 2	The Rough Wooing to the Glorious Revolution, 1540–1688	23
CHAPTER 3	Killiecrankie to the battle of Glenshiel, 1689–1719	50
CHAPTER 4	George Wade to the battle of Culloden, 1724–46	80
CHAPTER 5	The Roy Military Survey to Fort George, Ardersier, 1746–87	116
CHAPTER 6	The French Revolutionary Wars to the First World War, 1792–1918	154
CHAPTER 7	The Second World War to the present day, 1939–2018	184
	Guide to sources and further reading	219
	Index	223

PREFACE AND ACKNOWLEDGEMENTS

Warfare, attack and defence have shaped Scotland's history over the last six centuries. From the fifteenth to the seventeenth century, a prevailing ideology of English overlordship of Scotland created real threats and invasions through the Wars of the Rough Wooing in the 1540s, persistent violence on the debatable Scottish borderlands, and the Cromwellian occupation of Scotland in the mid seventeenth century. In the first half of the eighteenth century, the Jacobite uprisings, which in 1745 came close to toppling the British throne, led to a huge militarisation of Scotland, with new defences, forts and roads, and armies clashing in battle. Some of these defences were put to new uses by the late eighteenth and early nineteenth centuries, to counter the very real worries over French invasion, particularly on the east coast. By the twentieth century, defences and enemy threats had shifted in their focus again, with German sea and airborne attacks, particularly during the Second World War, followed by new fears over Russian military predominance in more recent decades.

This book uses six centuries of military mapping in Scotland to tell this story. It looks at Scotland's changing enemies over time, their motives and strategies, as well as the changing purposes and outcomes of warfare as revealed through maps. It aims to show and explain the variety of military maps produced for different purposes: fortification plans, reconnaissance mapping, battle plans, plans of military roads and routeways, tactical maps, enemy maps showing targets, as well as plans showing the construction of defences. We try to explain these within an international context, as well as through the particularities of the Scottish landscape and individual locations and people. Many of the military engineers who drafted maps were experts from overseas. They brought European skills, practices and styles to Scotland, where they proceeded to live and work, leaving their legacy in maps and fortifications. As well as illustrating plans, elevations and views, these military maps also show unrealised proposals and projected schemes – the paper military landscape that was never implemented. Many of these maps are both striking and attractive, and they have been selected for the particular stories they tell about attacking and defending Scotland.

We would like to record our thanks to many people who have assisted in the writing and production of this book. Hugh Andrew of Birlinn championed the idea of a book on military mapping several years ago, and we hope he will be pleased by what we have finally managed to deliver. Andrew Simmons and the staff at Birlinn have, as ever, been efficient, helpful and good-humoured, and we have been very lucky again to benefit from the graphic design skills of Mark Blackadder. We are grateful too to John Scally, National Librarian and Chief

Executive of the National Library of Scotland, who has continued to support and encourage these books which popularise and promote the Library's rich collections, as well as encourage research on them. More specifically, it is a pleasure to record our thanks to Vera Cebotari for her translation of the Russian military *spravka* of Aberdeen, and to Denis Rixson for helpful suggestions on John Hardyng and the enigmatic Monsieur de Bombelles at Killin in 1784. It will be clear from our 'Further reading' that this book is based on the substantial interest and scholarly attention that has been devoted to Scotland's military past. Military maps have long been a fascination of ours, and it is our hope that readers will come to appreciate Scotland's military mapping as an important way of understanding Scotland's history.

Image credits

Unless stated otherwise, all images are reproduced courtesy of the National Library of Scotland. We gratefully acknowledge all other sources for non-NLS map images in the captions for the respective individual figures.

Place names

Military maps of Scotland were nearly always made by outsiders, for outsiders, and place names are more likely to appear in English, French, Dutch, German, or even Russian Cyrillic, rather than reflecting any local Scots or Gaelic forms. Before the nineteenth century, there was no particular effort made to standardise names, and the same map-maker often cheerfully wrote different versions of the same place. Our general policy has been to use standard modern forms when we refer to a place, whilst in map titles or in quoting original sources, we use the original names. For modern Gaelic place names, the standard policy in the last decade has been to use bilingual English and Gaelic forms of names in the Western Isles/Na h-Eileanan an Iar. This follows Ordnance Survey's Gaelic Names Policy in association with the Ainmean-Àite na h-Alba (AÀA)/the Gaelic Place-Names of Scotland partnership. This reflects both local usage and official local government policy in the Western Isles/Na h-Eileanan an Iar. Elsewhere in Scotland, we follow the prevailing linguistic form, and do not give the bilingual name form as above.

CHAPTER ONE
INTRODUCTION

John Hardyng's maps of Scotland, originally drafted in the 1450s, were specifically intended to support an English invasion of Scotland, and they therefore provide an excellent starting point and introduction to the military maps of Scotland (fig. 1.1). Hardyng was brought up in the household of Henry Percy, Earl of Northumberland, and fought with Percy at the battle of Homildon Hill (1402), the siege of Cocklaw Castle (1403) and the battle of Shrewsbury (1403). After Percy's death in 1408, Hardyng entered the service of Sir Robert Umfraville, later Earl of Kyme; it was through Umfraville that Hardyng first came into contact with the English king, Henry V. Hardyng claimed that Henry had entrusted him with a special mission – to search out evidence for England's sovereignty over Scotland, and to plan the best route into Scotland for an invading English army. The pursuit of this mission dominated most of Hardyng's life.

Hardyng delivered to the Crown – first to Henry V in 1422, then to Henry VI in 1440 and 1457, and finally to Edward IV – a series of Scottish royal documents that purported to acknowledge English overlordship of Scotland. Although Hardyng claimed to have received these documents from Scotland, it is clear that the majority were entirely forged by Hardyng himself. He also made a detailed plan for the invasion of Scotland, which included a map of Scotland as well as a record of the distances between the major Scottish towns. These formed part of Hardyng's larger *Chronicle* of British history.

There are many similarities and continuities between Hardyng's maps of Scotland in the mid fifteenth century, the maps of the Rough Wooing made by the English in the sixteenth century, French invasion maps of Scotland in the early nineteenth century, and bombing and tactical reconnaissance maps by Germans and Russians in the twentieth century. All these maps are aligned with military strategy, supporting the practicalities of invasion and control of Scotland by a foreign aggressor. They all select particular features over others, picking out practical and useful topographic details to support military manoeuvres. For Hardyng, these were the main walled castles and towns forming the chief Scottish strongholds that the English army would need to focus on, and his supporting *Chronicle* describes these places as part of the itinerary that

Opposite. Советская Армия. Генеральный штаб, Великобритания, Шотландия, Абердин = Soviet Army, General Staff, *Great Britain, Scotland, Aberdeen (O-30)* (1986).

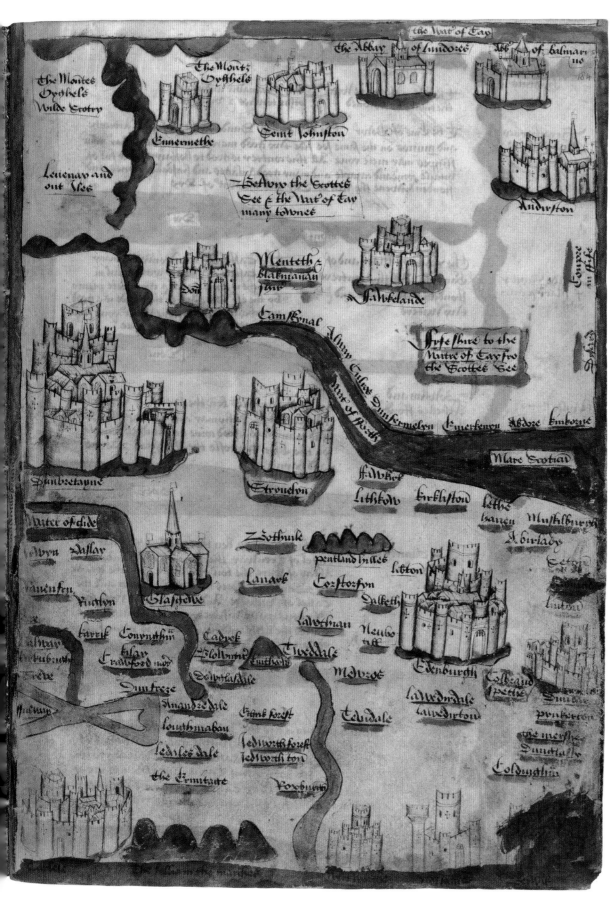

(a)

FIGURE 1.1

(a) Hardyng's map of Scotland shows seventy place names, from Carlisle and Berwick in the south to the River Tay in the north. The original map extends over three sides of two folios, with the second side (not included here) showing Scotland above the Tay, but without any drawings. The map picks out strongholds for the invading English army, including Berwick and Carlisle in the south, while above the Forth there are prominent drawings of Falkland, Perth (St Johnstone) and St Andrews. From these maps, it is evident that Hardyng's knowledge of Scotland north of the Tay was limited.

Hardyng's maps represent the earliest surviving military mapping of Scotland. This is the main part of the second of two versions of Hardyng's map, held in the Bodleian Library, Oxford. It accompanies the House of York's version of Hardyng's *Chronicle*, probably produced about twenty years after his death in 1464. Hardyng claimed that he was to report on the route the English army might take into Scotland and the towns on the east coast from which a fleet might supply the army with provisions and artillery. The route ran from Berwick to Dunbar, then

(b)

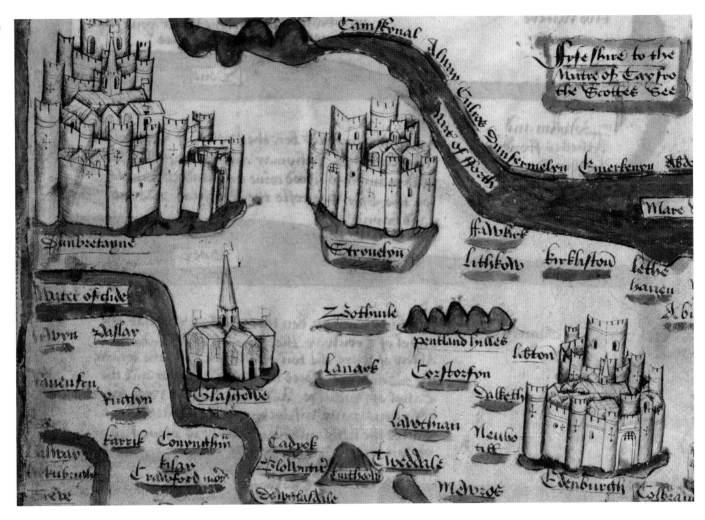

to Edinburgh and Stirling. From there the route ran north-east to Falkland, St Andrews, Perth, Dundee and Aberdeen, meeting the fleet at specific places *en route*.

> Nowe to expresse unto your noble grace
> The verie waye bothe by sea and land,
> With the distances of tounes and every myles space,
> Through the chefest parte of all Scotland,
> To conveigh an armie that ye maye take in hand
> Herafter shall folowe in as good ordre as I maye,
> The true discripcion, and distaunce of the waye.

Hardyng's main sources appear to have been the written accounts of previous royal invasions by Edward I, Richard II and Henry IV, supplemented by direct personal observation. Rather like Matthew Paris's map of the thirteenth century, Scotland is depicted almost as an island, perhaps reflecting the perceived political and ethnic division between Scotland and England, rather than any topographical conviction. There is less evidence from the names and outlines that Hardyng was influenced by *mappae mundi*, the Gough map of the fourteenth century, or the re-emerging Ptolemaic regional mapping.

(b) This detail from the same map shows central Scotland between the Clyde and the Forth, picking out the Pentland Hills and Tinto, and showing elaborate and distinctive architectural sketches of the castles and walled towns of Dumbarton, Glasgow, Stirling and Edinburgh.

Source: John Hardyng, 'Map of southern Scotland', from the *Chronicle* of John Hardyng (*c*.1470s). Courtesy of the Bodleian Library, University of Oxford.

the invading army might take. As for later military maps, Hardyng's map also has an important political and ideological role, in making accessible in the minds of the attacking military force the space of another country to be conquered. The significance of these maps lies less in their technical achievements than in their representation of key topographic intelligence on enemy territory to allow its subjugation and control.

This chapter looks at the main questions that guide the central narratives of this book. What value do maps have in providing an insight into Scotland's military past? What types of military map were made, and what were the driving forces behind their making? We also look at the people and institutions who produced these maps, why they were made and for whom. But first it might be useful to ask: what is a military map?

Types of military mapping and their purposes

Our definition of military maps, adapted from the current international History of Cartography project, is 'graphic representations that facilitate a spatial understanding of military things, concepts, conditions, processes or events in the human world'. This allows us to include plans, profiles, bird's-eye views, diagrams, charts, sketches and aerial photographs, recognising their common purposes. What these maps *do* and *mean* is more important than their physical attributes. This definition also allows us to include a broad range of maps – not just those created by military personnel, intended for a military audience, but also those by civilian personnel depicting military themes. The contrasts between these maps can be revealing, often highlighting the distinctiveness of maps made by military personnel, as well as the impact of military activities on civilian life. Within the context of selected examples in the chapters that follow, these categories can be usefully specified in further detail. Maps by military personnel can be categorised into four functions: reconnaissance; occupation; the construction or repair of military infrastructure such as forts, castles, roads and bridges; and the depiction of military encounters and

FIGURE 1.2

This detailed military map for fortifying the Anglo–Scottish border is a fine example of an imagined defensive work, never constructed. Christopher Dacre came from an old West Border family, whose support for the English in the battles of Flodden (1513) and Solway Moss (1542) had rewarded them with extensive lands around Lanercost, east of Carlisle. In the early 1580s, when the threat of war with Spain loomed, Dacre served on a special commission to survey defences along England's northern border. His solution was a new, grander Hadrian's Wall, running right across the country from the Solway to Wark on the Tweed, built to use and withstand modern artillery, rather like Berwick-upon-Tweed's recently constructed

INTRODUCTION

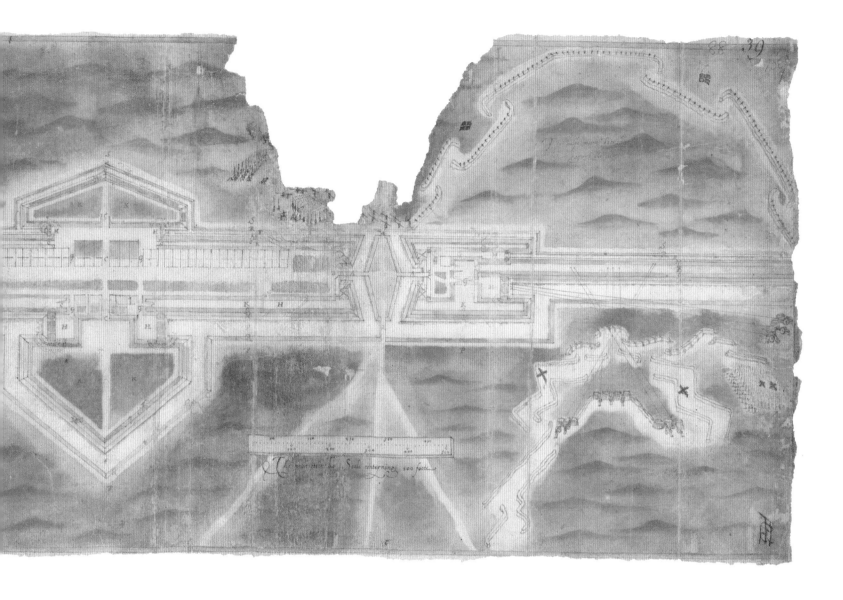

town walls. The proposed wall, 'The Forteficacions Royal', would be 60 feet thick with fortlets (or what he called 'sconces', deriving from the Dutch *skonse*) at every mile. The fortlets would be self-contained garrisoned villages, with artillery defences including angular bastions (bulwarks), casemates (flanking gun emplacements), ramparts and ditches. On the lower left of this elaborate plan for one of these 'sconces', Dacre provides a bird's-eye perspective of the main overhead plan, while on the right he shows multiple lines of offensive fire from the ramparts on an attacking army. Dacre is assumed to have sent this magnificent, if completely impractical, proposal to Queen Elizabeth I in the early 1580s, in a booklet entitled 'The Epystle to the Queen's Majestie', as a way of making England completely secure along its perilous northern frontier. Nothing came of the proposal, much to Dacre's bitter disappointment – the enormous costs of such a scheme were possibly a deterrent. Perhaps some already thought there would be cheaper and easier ways of creating peace between the English and the Scots.

Source: Christopher Dacre?, *Plan and Bird's-eye View of an 'Inskonce' (Sconce or Small Fort) for the Defence of the English Border with Scotland* (*c*.1583–84). Courtesy of The National Archives.

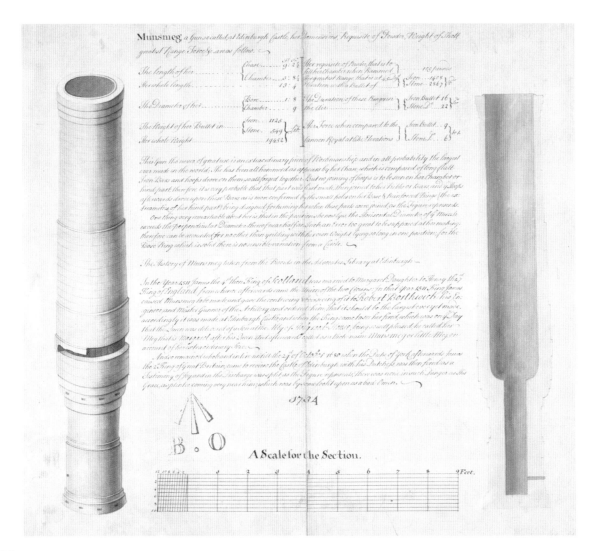

FIGURE 1.3

The famous Mons Meg siege gun, weighing six tonnes, was presented to James II of Scotland in 1457, and put straight to use in fighting the English in 1460 at Roxburgh, where James II lost his life. At the time, it was at the cutting edge of artillery technology, firing stones that weighed 380 lb, and with a barrel diameter of 20 inches, one of the largest guns in the world. As noted by the Board of Ordnance draughtsman on this drawing from 1734, 'this gun, tho' never of great use, is an extraordinary piece of Workmanship', and as illustrated, the manufacturer fused together longitudinal bars of iron, hooped with rings. However, as the gun could travel only three miles a day, hauled by a team of 100 men, moving it around to the next encounter was quite a challenge, and it was retired around 1550, superseded by lighter and more mobile cannons.

Thereafter it was fired for ceremonial purposes only, including on a visit by James, Duke of Albany and York, and later King James VII, in 1680, where (as illustrated) one of the iron rings burst, putting the cannon out of use. In 1754, following the Disarming Act of 1746, which demilitarised Scotland after the Jacobite risings, it was taken with other disused ordnance to the Tower of London. Spirited campaigns by Sir Walter Scott and the Society of Antiquaries of Scotland led to its return to the battlements of Edinburgh Castle in 1829. This drawing by an anonymous engineer could be an apprentice piece, exemplifying the versatility of the training in military draughtsmanship, including scales, measurements and a description.

Source: Anon., *Munsmeg, a Gun so called at Edinburgh Castle* (1734).

battles. It is also possible to see the life-cycle of maps relating to a particular defensive work, as well as proposals that were never executed (fig. 1.2). We have been keen, too, to try to show the different facets of military life, such as housing prisoners of war, constructing barracks, the need to secure water supplies and plan sanitation, and even the study of early monumental artillery (fig. 1.3). Other maps are useful in showing the internal social structures of military forces, the evacuation of civilians from towns, and the opposition to war and the military.

Warfare has been a primary driver behind mapping for centuries, and military maps of Scotland constitute important historical sources in their own right. The earliest town plans of Fort William, Inverness, Perth and Stirling (fig. 1.4), for example, were made by military engineers, and the first comprehensive systematic survey of mainland Scotland in the mid eighteenth century was a military one, under the superintendence of William Roy. Military maps are important, too, in illustrating many significant innovations in cartography and surveying, including the development of the scale plan, styles of colouring and relief presentation, accurate levelling, conformal projections, and the use of airborne cameras and sensors. The overhead plan, drawn to an accurate scale, has military roots in the early sixteenth century, when it was recognised as being the best way of showing projectiles, angles and distances around a fort. At this time, Italian military engineers brought this innovation with them to Britain, using it to promote new *trace italienne* artillery forts. Several Italian military experts were working in Scotland in the 1540s, as were some leading English engineers, who had recently learned continental engineering and cartographic practices at the English Pale of Calais. Continental styles of colouring features on military maps, as well as the use of hachures to show steepness of slope, were adopted in the eighteenth century as part of the Board of Ordnance house style in military mapping, and they are illustrated to perfection in the Roy Military Survey of Scotland (1747–55). These styles had a long-standing influence on later military and civilian maps. More recently, the development of conformal or orthomorphic projections, which preserve angles and bearings, was a great innovation for assisting in the accurate firing of artillery, and an important development of the First World War. In the twentieth century, developments in overhead sensors, including aerial photogrammetry, satellite imagery, LIDAR technology (using pulsed laser light to create three-dimensional representations) and drones (or unmanned aerial vehicles) were all driven by military requirements.

It is also useful to be clear about the military maps we exclude. Our geographic focus is Scotland, rather than the broader military history of the Scots overseas. We are not primarily interested in 'historical maps', that is, maps that look back on military sites or themes from an earlier period – for example, nineteenth- or twentieth-century maps showing the battle of Bannockburn or the Wars of Independence. The maps of Roman fortifications in Scotland by William Roy in his *Military Antiquities ...* of 1793 (fig. 1.5), or nineteenth-century Ordnance Survey maps illustrating (often for the first time) the many hillforts across Scotland, fall into the same category (fig. 1.6). These maps certainly have their uses, but they inevitably reflect to a greater or lesser extent the values of the time period

FIGURE 1.4

(*Overleaf*) This clear and striking military plan is our earliest surviving detailed map of Stirling. It allows us to see the town with a military eye, picking out many details of the castle fortifications and its surrounding topography, while also giving good information on the town, including the church, hospital, major market places and the earliest clear depiction of the sixteenth-century walls. It shows the results of the 'New work & barracks by Mr Dury', the major refortification programme carried out in 1708–14 (fig. 3.6), and the French spur at the top, which gave covering fire along the front. It marks all the significant buildings within the castle (A), and includes Argyll's Lodging (L), Mar's Walk (M), the bowling green (R) by the hospital (O), and even the gallows to the south-east (S). The plan superbly illustrates the Board of Ordnance house style, picking out man-made features in red, and distinguishing enclosed and open ground, gardens and types of tree cover.

Source: John Laye, *A Plan of the Town and Castle of Sterling* (1725).

PLAN of the ENVIRONS of the EILDON HILLS on the SOUTH BANK of the TWEED at the bottom of which it is supposed the TRIMONTIUM of the ROMANS was situated.

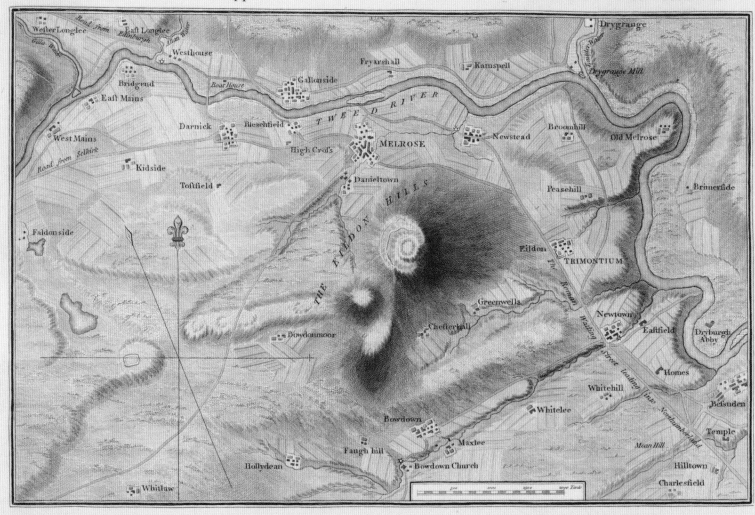

VIEW of the EILDON HILLS as they appear from the SOUTH EAST, the QUARTER from which the ROMANS would first discover and approach them.

INTRODUCTION

FIGURE 1.5

(*Opposite*) William Roy was the first person to correctly link the Roman fort of Newstead and the Eildon Hills with Trimontium in Ptolemy's *Geography*. This attractive view of the Eildon Hills, near Melrose, was published in Roy's *Military Antiquities of the Romans in North Britain* (1793), a year after Roy died. Roy was a keen antiquarian, and when visiting his mother near Carluke in the summers of the 1760s, he undertook detailed archaeological investigations of Roman sites in Scotland. Roy wished his antiquarian hobby would be regarded as 'the lucubration of his leisure hours, [rather] than as tending to any great utility', but he was well aware of the close connections between the Roman conquest of Scotland and the work of later military commanders, as well as the military value of studying much earlier military features in the landscape. As he noted:

> The nature of a country will always, in a great degree, determine the general principles upon which every war there must be conducted. In the course of many years, a morassy country may be drained; one that was originally covered with wood, may be laid open; or an open country may be afterwards inclosed: yet while the ranges of mountains, the long extended valleys, and remarkable rivers, continue the same, the reasons of war cannot essentially change. Hence it will appear evident, that what, with regard to situation, was an advantageous post when the Romans were carrying on their military operations in Britain, must, in all essential respects, continue to be a good one now.

Source: William Roy, 'Plan of the Environs of the Eildon Hills on the South Bank of the Tweed . . .', from *Military Antiquities of the Romans in North Britain* (1793).

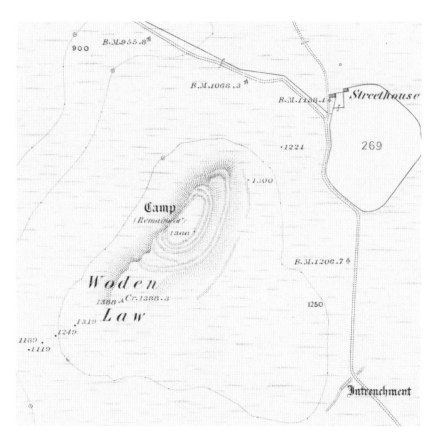

FIGURE 1.6

(*Above*) The impressive hillfort of Woden Law, with commanding views, particularly to the north and west, lies just north of the current Anglo–Scottish border across the Cheviots, eight miles south-east of Jedburgh. This highly strategic site lies adjacent to the main Roman north–south route of Dere Street, a once crucial link between York and the Firth of Forth. Dere Street probably fell into disrepair after Agricola's withdrawal from Britain around AD 100, but it was reconstructed during the Antonine reoccupation of the second century. The fort was certainly in use by native Britons before the Roman occupation of Scotland, and, although it was abandoned during this time, it was refortified in the third century AD. It is possible that some of the ditches are Roman siegeworks.

The many Iron Age and other hillforts scattered across Scotland are testimony to the need for defensive structures at this time, whether for status or real protection. Often the earliest detailed cartographic record of them is by Ordnance Survey, in their first edition mapping of Scotland from 1843 to 1882. While at this time Ordnance Survey may not impress us with their archaeological knowledge of structures (which improved rapidly with aerial photography and following the appointment of their first Archaeology Officer, O. G. S. Crawford, in 1920), their maps are often the best for showing the detail of these earthworks, before later alterations on the ground and changes in map style.

Source: Ordnance Survey, Six-Inch to the mile, *Roxburghshire, Sheet XXVIII* (surveyed 1859, published 1863).

of their creation rather than the military topic depicted; our focus is not primarily historiography. It is for this reason, too, that our military history through maps is very much a history of the last six centuries – the time period for which maps of Scotland survive today. We also recognise that our coverage is necessarily highly selective, and drawn particularly from the collections of the National Library of Scotland. Another significant exclusion is maps that are still in copyright, particularly those published in the last fifty to seventy years. Our hope is to encourage further interest, to illustrate broader themes and outlines, aiming for breadth, balance and variety, rather than an attempt to be comprehensive.

Mapping the military history of Scotland

Our central purpose in this book is to illustrate the military history of Scotland from the perspective of maps and map-making. The maps can be read at many levels to provide important insights into Scotland's military past. At one level, these maps illustrate the dramatic changes in Scotland's 'enemies' or attacking military forces over time. During the English invasion of Scotland as part of the War of the Rough Wooing in the 1540s, Scotland was mapped for offensive purposes. Following the Union of the Crowns in 1603, and especially the Act of Union of 1707, Scotland's army and military engineering infrastructure were increasingly absorbed into British institutions, and Scotland was increasingly drawn into British conflicts. Scotland suffered military attacks, for example, during the American Revolutionary War (1775–83), the French Revolutionary and Napoleonic Wars (1792–1815), and the First and Second World Wars, but as part of Great Britain. Scotland's own unique culture and history have also resulted in warfare, particularly through the Jacobite rebellions and uprisings in the period from 1689 to 1746. This period saw significant military mapping by the British Board of Ordnance, with upgrades to existing Scottish castles, new forts and roads, and related defensive infrastructure put in across Scotland (fig. 1.7). The enforced pacification of the Highlands following the battle of Culloden in 1746 was achieved partly through maps. During the twentieth century, the growing internationalisation of warfare saw the threats to Scotland from further afield steadily grow, with impressively detailed German and Russian military mapping of Scotland, as well as the siting of important British military bases and training grounds within Scotland. The distinctiveness of Scottish military mapping reflects the uniqueness of Scottish history.

At another level, these maps show widespread transformations in warfare over the centuries. An early focus on traditional castles, with relatively small armies and a limited artillery range, changed rapidly during the sixteenth century, as Scotland faced rivals with much larger armies, new firearms and vastly superior mobile siege guns with a greater destructive potential. The transition in Scotland from blood feud to more centralised state control of military forces in the late sixteenth century was a driving force behind map-making, and the maps themselves also revealed these changes. For example, from the late sixteenth century, the earlier, more geographically dispersed scattering of military strongholds, with a relatively localised range, gave way to a smaller number of state-controlled sites. The earliest detailed graphic depictions of many of Scotland's tower houses and castles by Timothy Pont show lofty structures, but these were designed to deal with localised conflict, not great artillery fire. The need to protect Scotland's castles from Jacobite attack from the early eighteenth century led to major upgrades to existing castles, new innovative defensible barracks and several imposing, completely new forts, adopting the latest thinking in military engineering. At the same time, a network of military roads was constructed across Scotland, allowing men and munitions to be rapidly moved around the whole country. As the size of armies grew, particularly from the late eighteenth century, so did the need for barrack accommodation. The Napoleonic Wars also saw major coastal fortifications, through martello towers and batteries protecting the major harbours. The twentieth century has seen even greater changes with innovations such as the tank, submarine and aeroplane rendering many former fortifications obsolete. In their turn, concerns about naval shipping, invasion beaches

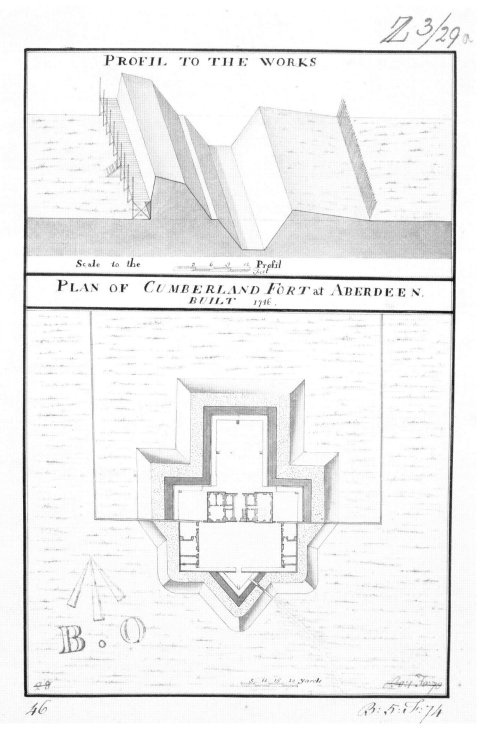

FIGURE 1.7

Robert Gordon's Hospital or College was constructed in 1732, to designs by William Adam, following a generous bequest from the Aberdeen merchant Robert Gordon, who had died the previous year. The original building was still empty when the Duke of Cumberland arrived in Aberdeen on 27 February 1746, and requisitioned it as his military headquarters. This plan shows the substantial earthworks, with a ditch and V-shaped ravelin (an external triangular fortification), which were intended to provide defences to the building from artillery fire. In practice this never came, perhaps for the best, as the works were not completed, and the duke and his troops marched out on 8 April to defeat the Jacobites later that month at Culloden. While the troops were garrisoned there, a considerable amount of damage was also done to the interior and the governors put in a claim for the cost of repairs to the king, which was paid in October 1747. The school eventually opened to pupils in July 1750.

Source: Anon., *Plan of Cumberland Fort at Aberdeen Built 1746* (1746).

and coastal defences up to and during the Second World War were rendered largely irrelevant with the advent of nuclear missiles from the 1950s. Many Scottish fortifications were subsequently abandoned, sold for new development or levelled, except those gaining statutory protection. There are few documentary sources as useful as maps for illustrating these changes.

A third level is how maps depict the changing geography of military activity in Scotland over time. The English invasions of Scotland in the sixteenth century, for example, focused on the Borders and major towns, and scarcely affected the Highlands. The significant disturbances of the Wars of the Three Kingdoms (1639–53) and the Cromwellian occupation of Scotland had a wider geographical impact, but still centred on particular places. During the Jacobite uprisings, and especially after 1715, the focus of attention moved to the Highlands, with new military roads, barracks and forts. Growing concerns of attack from continental Europe saw significant defences along the east coast in the early nineteenth century and again in the twentieth century. The threat of airborne bombing as well as seaborne invasion in the Second World War had a much wider impact on the whole population. In the twentieth century, Scotland's perceived remoteness from major centres of British population resulted in its militarisation for training purposes, military ranges and airfields. The Northern and Western Isles/Na h-Eileanan an Iar have become major military bases for missiles or listening purposes, whilst the Clyde has become the largest nuclear submarine base in Britain. More recent British military involvement in many overseas conflicts, allied to sales of British armaments around the world, has exacerbated a range of newer threats, including terrorism.

At a fourth level, and perhaps of most importance, these military maps all illustrate both more and less than the landscape they purport to show. The art and artistry of cartography, the conscious and unconscious selections and exclusions that all map-makers need to make, as well as the deliberate embellishments and distortions for particular purposes, all allow an infinite range of different ways of seeing the same world. The different maps produced by defenders or attackers of the same place, or by the victors or losers of the same battle, clearly reflect this point. Military maps were essential for boosting morale and imbuing their users with confidence in their superior topographic intelligence. They needed to communicate essential military or tactical information, and quickly. Reducing or removing the map of other, non-military clutter and providing much-needed clarity was essential in times of war. The power of maps in encouraging their users to think and act in particular ways – attacking, conquering or even killing other people – arguably has its most dramatic and destructive effects through maps depicting military activity (fig. 1.8). Military maps encourage violence, whether through direct attack or invasion, or through engendering such activities. More generally, the military institutions within which maps were made, with their clear hierarchies, codes and procedures, not only created their own 'house styles', but also encouraged particular ways of representing the world. At a very simple level, a humble fortification proposal would need to impress its readers, who controlled funding, that the construction project was worthwhile, or that the existing fort was beyond repair (see fig. 1.9). Collectively, the institutions that produced military maps served governments, states and nations, and their maps reflect this loyalty. For civilian topographic mapping of military features meanwhile, the 'cartographic silences' of those features excluded from view is very obviously illustrated by state censorship, which has rapidly risen and fallen at particular times in the past (see fig. 7.3). These maps all represent the world, but they do not do so in a simple fashion.

Who produced military maps, why and for whom?

Most of the military maps illustrated in this book reflect the work of military engineers, many of whom were also skilled draughtsmen who drew the maps themselves. Many of these engineers were not from Scotland, or even Great Britain, and their work, including their maps, can only be properly understood through an international perspective. Over the centuries, the main centres of military engineering expertise shifted in continental Europe. We find Italian expertise in the sixteenth century overtaken by Dutch and then French theory and practice by the later seventeenth century. The best engineers were polymaths – men who could combine mathematical skills with trigonometry and geometry to measure heights, distances and volumes, together with architectural training to design and provide estimates for constructing defensive works. Other essential skills included a knowledge of ballistics, gunpowder and artillery to plan and conduct siege operations, as well as surveying and draughting skills to compile and produce maps of all aspects of the military landscape. There was a slow but steady institutionalisation of military engineers over time, which we see in Scotland in the seventeenth century, and Great Britain especially in the eighteenth century. These institutions changed their name over time – the work of the Board of Ordnance in the eighteenth century was effectively replaced by the War Office by the late nineteenth century – but their role and association with military cartography and intelligence remained throughout.

Another reason why an international perspective is essential for understanding military maps is that military engineers were highly mobile and often willing to work for different states as opportunities arose. Lewis Petit, who drafted early plans of several castles in the north-west Highlands (fig. 1.10), as well as the earliest town plans of Perth and Inverness (chapter 3), came from Caen in Normandy. He left France soon after the revocation of the Edict of Nantes in 1685, which curtailed civil rights and freedom to worship for French Huguenots. He gained valuable practical military engineering expertise in Spain, generally fighting French forces in the Wars of the Spanish Succession (1701–14). For a brief time from 1707, he was placed in command of the new British possession of Minorca. Owing to his first-hand military experience in Europe, he was chosen to advise on Scottish defences in and around Fort William in late 1714, and led the artillery train into Scotland to crush the Jacobite Rising of 1715.

As a result of their expertise and travel, often in conflict zones, the best military engineers could pass on very useful topographic intelligence to whoever they chose to work for. John Elder and Nicolas de Nicolay, who both compiled maps of Scotland in the sixteenth century, were spies and happy to transmit topographic information to England and France. Thomas Petit, who compiled a military map of the English 'pale' in south-west Scotland in 1547 (chapter 2), was surveyor of Calais, working for the English, and he was briefly captured in 1548 whilst in Scotland. There is a nice account, too, by the French geologist and traveller Barthélemy Faujas de Saint-Fond (1741–1819) of his meeting with a fellow Frenchman, M. de Bombelles, at Killin in 1784. '"I am Bombelles" said he; "I travel, like yourself, for pleasure and instruction. I am now on my way to Port Patrick . . ."' However, Faujas de Saint-Fond quickly felt things were not quite so straightforward. 'From the career which M. de Bombelles followed, as well as from a number of military and other charts which he had along with him, I judged that diplomacy and politics were more to his taste than the natural sciences and the arts, and that he was probably charged with some particular mission very foreign from the object of my studies.'

Personal connections and family ties were often very important for the transmission of engineering and related cartographic expertise. The Irish Cromwellian Colonel Thomas Blood became a court favourite of Charles II, surprisingly following an unsuccessful attempt to steal the Crown jewels from the Tower of London in 1671. The bid had been foiled by the unexpected arrival of another military engineer, Talbot Edwards, son of the custodian of the regalia. Thomas Blood's son was Holcroft Blood, Captain of the Pioneers, who worked under Sir Martin Beckmann, First Engineer of Great Britain, from 1685, whilst Talbot Edwards went on to become Second Engineer of Great Britain, and Chief Engineer of Barbados and the Leeward Islands. Whilst in Barbados, he looked after his wife's nephew, William Skinner, who had been orphaned at a young age, but who clearly learned much from his adopted father. Less than a month after Edwards died in 1719, Skinner joined the Board of Ordnance as a Practitioner Engineer, going on to become Chief Engineer of Great Britain from 1757, and attaining the rank of Lieutenant-General.

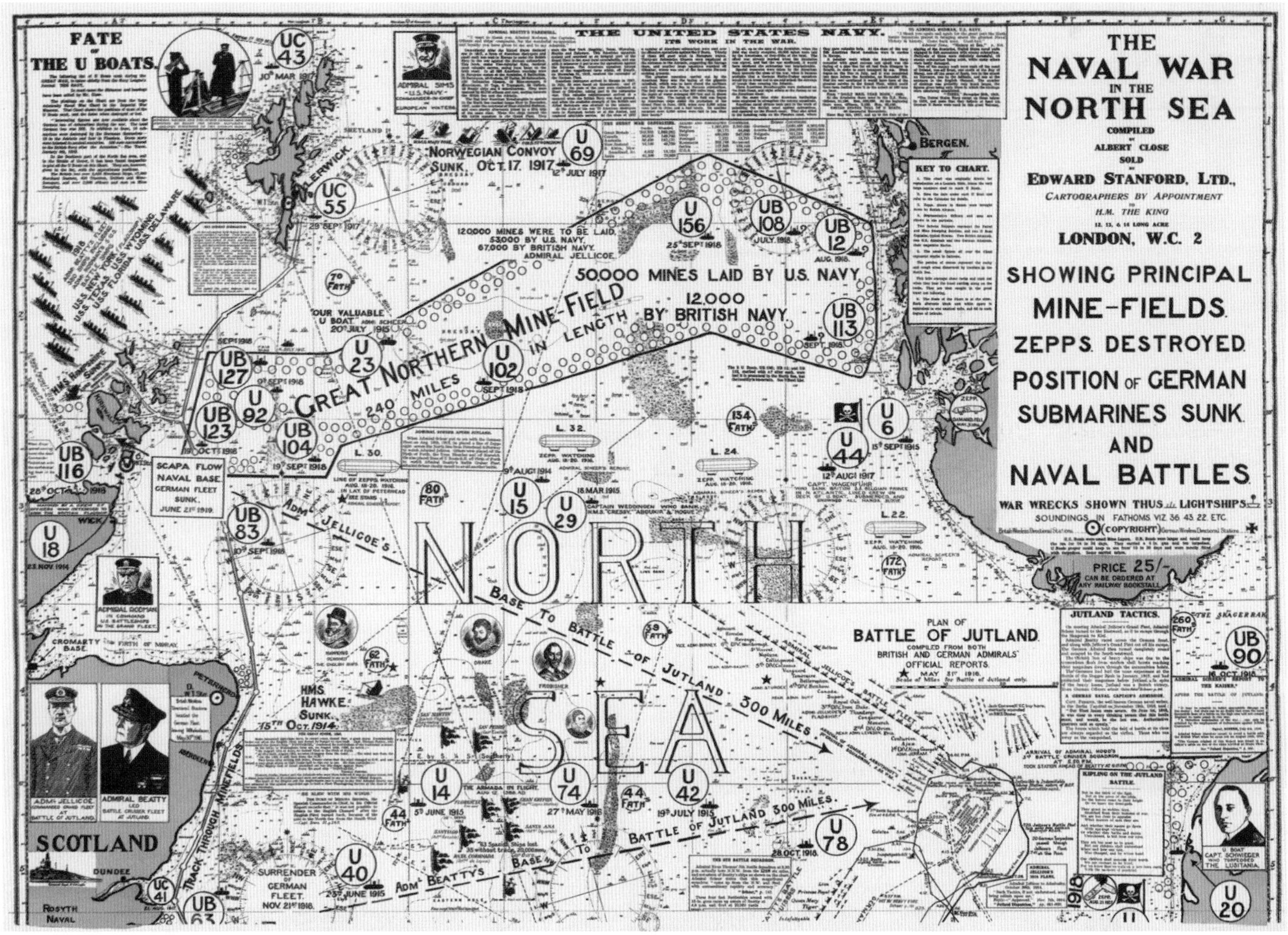

FIGURE 1.8

(a) (*Above*) This action-packed map celebrates Allied naval successes in the First World War in the North Sea. It is far from being a balanced or straightforward depiction of naval action, focusing very much on German ships and U-boats sunk, the successes of British minefields, heroic British and US Admirals, and the German surrender. The map was originally drawn for reproduction on a lantern slide, which explains the bold features and large fonts.

Albert Close drafted a number of maps from the early 1900s through to the 1930s, many of which were published by Edward Stanford, the London map publisher and dealer. Close's initial work focused on maps of British waters for deep-sea fishermen, but he went on to draw a few quite distinctive maps of the First World War, including this map. He seemed to be unaware that his cheerful copying of British Admiralty charts over three decades for his main topographic content infringed their copyright, for which he was prosecuted in 1931. He was also a keen member of the Protestant Truth Society, and turned in later years to draft quite virulently anti-Catholic maps and written publications, including *The Protestant Historical Map of Britain from Wickliffe to the Defeat of the Romanized Prayer Book AD.1374–1927* (1932).

For Close, these wartime maps carried quite clear British imperial, anti-German as well as anti-Catholic messages. The contemporary detail is also complicated by showing the passage of the English Armada through the North Sea in 1588. As Close explains in a note, the great storm in 1588 was regarded as divine Providence – 'The Mighty God of Israel stretched out His finger against them.' He went on to conclude 'The heart of the Empire and the Anglo-Saxon race is proud to-day that Admiral Beatty was not ashamed in this cynical, superficial age to hold a Fleet Thanksgiving Service at the close of the German Surrender.'

(b) (*Opposite*) A detail of the map showing the North Sea east of Scotland.

Source: Albert Close, *The Naval War in the North Sea: Showing Principal Mine-Fields, Zepps Destroyed, Position of German Submarines Sunk and Naval Battles* (1922).

FIGURE 1.9

This plan records the sorry state of the former Fort George in Inverness, showing how beautiful your blown-up castle could be made to appear through the skills of an accomplished draughtsman like Charles Tarrant. The commander at Fort George quickly surrendered to the Jacobites on 21 February 1746, unable to check the progress of a Jacobite mine beneath the ramparts. The Jacobites then blew up and burnt as much as they could, attempting to render the fort useless in case it fell back into Hanoverian hands. This plan shows how successful they had been at this, recording in detail 'the Several parts of the Fort, that the Rebels destroy'd by their Mines'. On the left is the 'The Governors House intirely in Ruins', and on the right 'The Old Castle, Formerly usd as Officers Barracks, part Blown up'. To the top right are the 'Ruins of the Chappel', while at the bottom are 'Barracks for Soldiers, the Walls ruin'd, the Roofs and Floors, Burnt'. By 1750, all efforts were focused on constructing a new, much larger replacement Fort George at Ardersier, which required detailed justification as to why the former fort was beyond repair. In the following decades, the former castle was no doubt plundered of anything useful, and in 1791, the *Old Statistical Account* recorded that nothing now remained of the castle, and the site was filled with rubbish. It was not until 1834 that the foundation stone was laid of the new County Buildings and Court House now visible on this site, built in 1836–43.

Source: Charles Tarrant, *Plan of Fort George at Inverness, Shewing its present Condition* (1750).

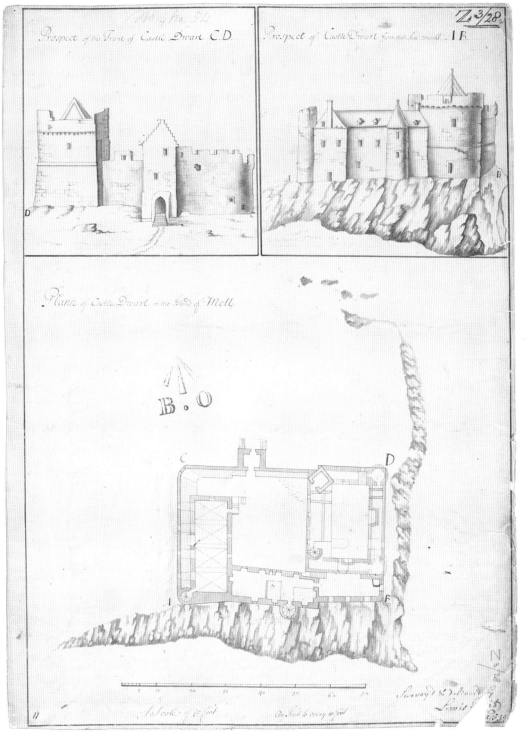

FIGURE 1.10

Castle Duart commands a superb strategic location on the south-east of Mull, overlooking the approaches up Loch Linnhe to Fort William and the Sound of Mull. This ancient fortification had been remodelled many times over the years, while also changing hands from MacDougalls to MacLean ownership in the thirteenth century, and then to the Campbells in the 1670s. It was garrisoned some time after 1708 by the Hanoverian government, along with a scattering of other outposts around Fort William, and in 1714 fifty soldiers were stationed here.

The Board of Ordnance engineer Lewis Petit was commissioned to survey Duart between September and November 1714, along with various other defences in and around Fort William. His *Prospect of the Front of Castle Duart* to the upper left, looks at the building and its entranceway from the south-west; the companion sketch shows the view from the Sound of Mull to the north-east. Petit's profiles and plan confirm that the castle had strong outer ramparts, but its main tower house on the north-west side was partly ruinous by this time. The soldiers were accommodated in barrack rooms on the south-east side. Petit's work is utilitarian and informative, not recommending changes but recording the castle's essential military properties. Duart, along with other garrisoned castles in this area, was more provocative to the Jacobite clans than effective, and it was in a more ruinous state when later recorded by Paul Sandby in 1748.

Source: Lewis Petit, *Plan of Castle Dwart in the Island of Moll*; *Prospect of the Front of Castle Dwart CD*; *Prospect of Castle Dwart from the Sea markt AB* (1714).

Personal connections are also crucial in understanding how the young William Roy was placed in charge of the Military Survey of Scotland (1747–55), and who assisted him. Roy was chosen for the post by David Watson, Deputy Quartermaster-General in Scotland from 1747, who also had close links with the powerful Dundas family of Arniston. Watson's sister Elizabeth married Robert Dundas (1685–1753), the third Lord Arniston, and Watson became a close friend of his son, Robie Dundas (1713–87), who became fourth Lord Arniston from 1753. When Elizabeth died in 1733, Robert married Ann Gordon, whose family not only owned extensive estates in north Lanarkshire, but also estates where William Roy's father and grandfather worked as land stewards, and it is very likely that Watson met the young William Roy through these family connections. The Dundas family was a major legal and political dynasty; Robie's brother, Henry, First Viscount Melville from 1802, had become one of the most powerful men in Scotland by the late eighteenth century. Following the defeat of the 1745 Rising, the Dundases helped to instigate numerous measures to suppress Jacobitism, and map-making played an important role in this. The Dundases supported Roy's work on the Military Survey financially through David Watson, and both William and David Dundas were directly employed as cadet engineers working with Roy (chapter 5).

Chapter structure

The chapters are structured chronologically, as the specific combinations of warfare and combatants reflected in the maps are easiest to understand within their particular time periods. Whilst this does split up particular locations, people and the primary purposes behind the mapping, we try to make obvious links between these themes across the chapters. Owing to the uneven distribution of military activity and maps over time, our chronological chunks are necessarily uneven. We look initially at the War of the Rough Wooing in the 1540s, the Cromwellian citadels and forts of the 1650s, and military engineering at the time of the Restoration. The Jacobite risings, from 1689 to 1746, take up two chapters, and the aftermath and responses take up a third; nearly half of the book is devoted to the sumptuous and detailed Board of Ordnance manuscript maps and plans, which portray the various attempts to resist and respond to the Jacobite threat. These plans are rightly important, but we hope that our longer history of military mapping in the centuries before and after the Jacobite risings allows their special characteristics to be better appreciated and understood, as well as showing some of their continuities with earlier and later mapping. From the time of the Act of Union in 1707, Scotland's defences and those administering them changed; so, too, did its military threats. As part of Great Britain, Scotland's military needs and defences in terms of funding, attention and map-making also had to compete with ever-expanding military requirements across a growing British Empire. Our final two chapters look at the impact on Scotland of the French Revolutionary Wars, the First World War and the Second World War, with increasingly globalised military threats and new types of warfare and mapping. In these centuries, we also see a major shift in defences, from landward fortifications to naval and then air power, with a steady broadening of the military landscape and those included in it.

FIGURE 1.11

(*Opposite*) Ordnance Survey, Scotland, 1:253,440, Royal Air Force Edition, Sheet 3 – The Forth & Tay. (1939).

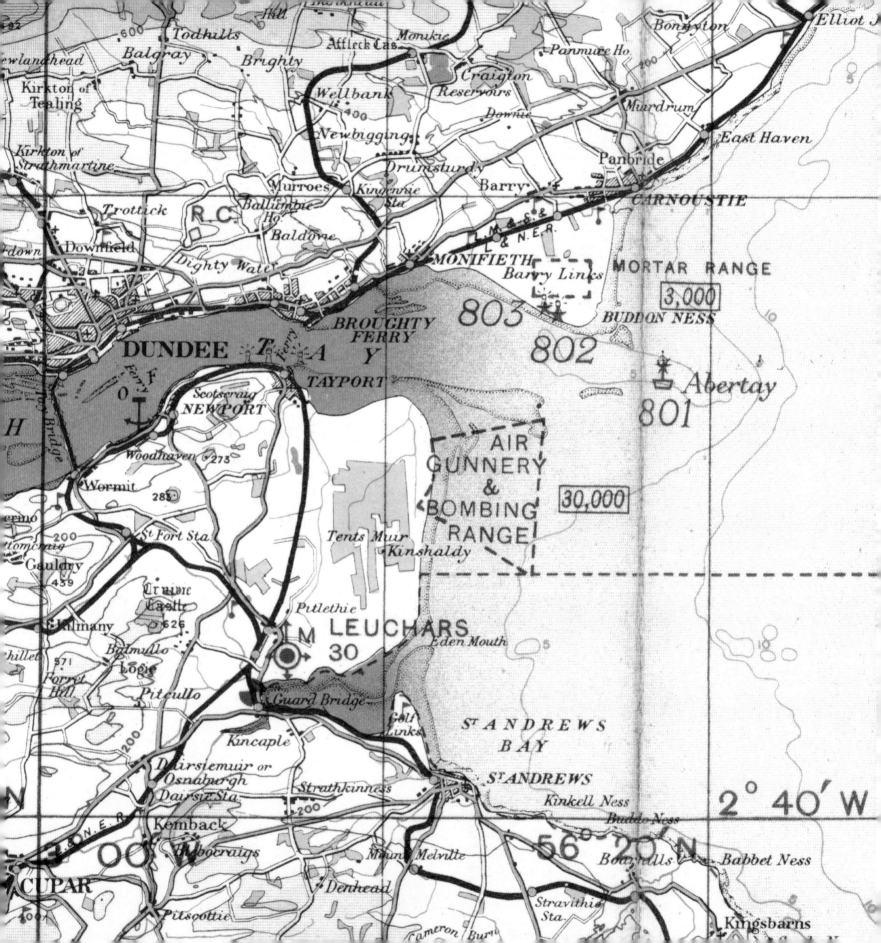

CHAPTER TWO
THE ROUGH WOOING TO THE GLORIOUS REVOLUTION, 1540–1688

On 12 June 1540, King James V set sail from Leith on a circumnavigation of the Scottish mainland. The voyage took the best part of a year, visiting Orkney, Skye, Lewis/Leòdhas, Wester Ross and Kintail, before disembarking at Dumbarton. The voyage was important for both military and political reasons: as a show of strength to 'daunt' the Isles and, like his father before him, for James to meet with various clan chiefs to receive their submission to his authority.

What is important for our purposes is that the navigation was assisted by a rutter or 'routier' – a detailed written set of distances and directions, with hazards and notable landmarks described and named – compiled, it is assumed, by the expert Scots pilot, Alexander Lyndsay. A map, now lost, is also referred to. The subsequent 'voyage' or travels of Lindsay's rutter tell us much about the importance of geographic and topographic information in sixteenth-century Scotland – who compiled it, who needed it and what form it took.

The rutter makes its next appearance seven years later, in 1547, now in French hands, assisting with an expedition to lift the siege of the castle in St Andrews. But how had the French got their hands on it? The answer lies partly with the admiral, general and politician John Dudley (1502?–53), later 1st Duke of Northumberland, and his attempts to trade military secrets with the French foreign agent Nicolas de Nicolay (1517–83). Nicolay had a long and active life, at various times working as an artist, courtier, foreign agent, traveller, expert on military fortification, navigator, geographer and cartographer. In 1546, he met John Dudley when the latter was in Calais attempting to broker peace negotiations between England and France, and returned to England with him. Dudley, who had recently distinguished himself at the battle of Solway Moss (1542), was perhaps the second most powerful man in England at this time, after Edward Seymour, Earl of Hertford. Dudley had managed to obtain a copy of Lyndsay's rutter, probably during his time as Warden of the Scottish Marches in 1542. As Nicolay later wrote 'in order to bring me more fully into his designs, he communicated to me a little book written by hand in the Scottish language . . . together with the sea-chart roughly made . . . I was unwilling to part with it without retaining a copy'. In a letter to the Duke of Joyeuse, Nicolay made it clear that Dudley's gift of the Scottish rutter was very much in recompense for Nicolay's 'Chart and

Opposite. Thomas Geminus alias Lambrechts, *The Englishe Victore agaynste the Schottes by Muskelbroghe* (1547).

Geographical Description of the Island and Kingdom of England, in which I had observed several noteworthy and uncommon matters'.

Nicolay wasted no time in getting Lyndsay's text translated into French and presenting it to King Henri II. Henri immediately sent it to military expert Leon Strozzi, who led the expedition to St Andrews. Nicolay also took part in the expedition. The fleet left Rouen in late June 1547, and took St Andrews Castle on 31 July 1547 (fig. 2.1). The French would also have found the rutter useful in their subsequent alliance with Scotland from 1548, and their occupation of Leith until 1560 (fig. 2.2). When the rutter was finally released for wider public consumption in Nicolay's printed map of 1583, it became the most geographically accurate outline of Scotland for the next two centuries, testifying to the extraordinarily high quality of Lyndsay's original work (fig. 2.3).

These events serve as a useful introduction to several underlying themes of military maps in sixteenth-century Scotland. Firstly, this was an era when graphic works, such as maps or plans, were still relatively novel compared with the primacy of written information – the rutter, for example, was a textual work long before it was printed as a map. Second, it is clear that during the sixteenth century there was a rapidly growing appreciation of the value of these new maps for military purposes. Maps and geographical texts were increasingly sought-after for their military value and traded as military secrets. Third, the survival rate of sixteenth-century maps is low, and hand-drawn manuscript maps are even more scarce; the resulting printed map and copies of the rutter are the best we have, as the original text and the related map that it described were lost long ago. Finally, although England was Scotland's main 'enemy', as reflected in the military maps of this era, the shifting sands of sixteenth-century politics and religion created a more complex picture, and the same geographical information was reused by different sides for different purposes.

This chapter will look first at the Rough Wooing, 1544–50, and the dramatic impact it had on Scotland's army and defences, as well as driving the creation of military maps. Military map-making continued in the following decade with attempts to create order in the Debatable Land on the Anglo–Scottish border, as well as to lift the siege of Leith in 1560. We then look briefly at the nationwide topographic surveys of Timothy Pont and their publication in 1654 by Joan Blaeu, and the value of these non-military maps for understanding Scotland's military infrastructure at this time. The Cromwellian invasion of Scotland and the construction of Protectorate citadels in the 1650s forms a fourth theme, before we conclude briefly with a look at the early rise of a military engineering institution in Scotland through the work of Captain John Slezer.

The 'Rough Wooing'

In 1543, the Scots rejected Henry VIII's attempt to broker a dynastic union between England and Scotland through the marriage of his son, the future King Edward VI, to the infant Mary, Queen of Scots. Scotland then suffered a series of English invasions, with military occupation, garrisoned towns and new English fortifications in the country – a period that came to be known as the 'Rough Wooing'. In an effort to force the Scots' hand, southern Scotland was initially invaded in 1544 through an amphibious operation commanded by Edward Seymour, Earl of Hertford, his massive army supported by a naval fleet. Edinburgh was attacked and badly burnt, with the English also plundering several other towns and villages in their retreat from Scotland. In 1547, Seymour's army returned and there was a bloody and decisive English victory at Pinkie Cleugh near Musselburgh. The English strategy involved fomenting rebellion in the Highlands and creating an English 'pale' in the Lowlands peopled by those known as 'assured Scots', who took an oath of loyalty to the English Crown. The Scots looked to France for assistance and, over the next decade, French troops in Scotland not only helped resist English forces, but also brought an influx of new continental engineering expertise. However, with the betrothal of Mary to the French dauphin in August 1548, the primary objective of the Rough Wooing had failed. Following the French declaration of war on England in August 1549, the loss of the English garrison in Haddington in

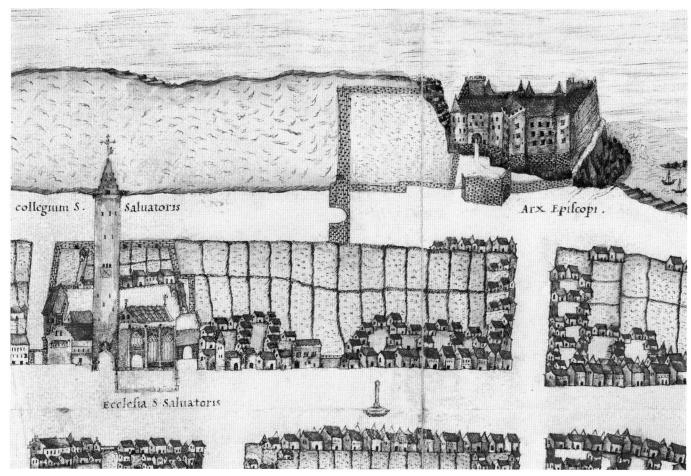

FIGURE 2.1

This detail from a striking bird's-eye view of St Andrews, attributed to John Geddie, shows the state of the castle and St Salvator's College at around 1580. Cardinal James Beaton, Lord Chancellor of Scotland from 1515 and Archbishop of St Andrews from 1521, actively refortified the castle, strengthening the south and west walls and redesigning the fore tower (to the right of the entrance) to withstand attack by gunpowder artillery. Beaton and his nephew, David, who succeeded him in 1539, were vigorous opponents of Protestant ideas circulating in Scotland, and actively persecuted Reformers, including Patrick Hamilton (1528) and George Wishart (1546). David Beaton's ties with France and opposition to Henry VIII's attempts to control Scotland added to his enemies. In May 1546, with the backing of the English, a group of Fife lairds, disguised as stonemasons, gained access to the castle and murdered Beaton – his naked body was subsequently hung in a pair of bed sheets from the castle walls.

The castle was besieged on the orders of Regent Arran – a long and difficult siege, lasting almost a year. The besiegers tunnelled towards the walls, but the defenders countermined and took their tunnel – both tunnels can still be seen today. The stalemate was broken when a French fleet arrived in July 1548 to support Arran, carrying out a devastating artillery bombardment from the sea, aided by the famous military expert Leon Strozzi, Prior of Capua. Guns were also fired at the castle from the towers of St Salvator's and the cathedral, soon rendering the castle indefensible, with the east range largely destroyed. A steeple was added to St Salvator's to prevent its later use by artillery. Beaton's successor, John Hamilton (the illegitimate half-brother of Regent Arran), relocated his episcopal state apartment to the south range, but he shared a similar fate to his predecessors, and was hanged in 1571, having been accused of complicity in the murders of Henry Stuart, Lord Darnley and the Regent Moray. The castle continued to be occupied by later archbishops or their constables until the mid seventeenth century.

Source: John Geddie?, *S. Andre sive Andreapolis Scotiae Universitas Metropolitana* (c.1580).

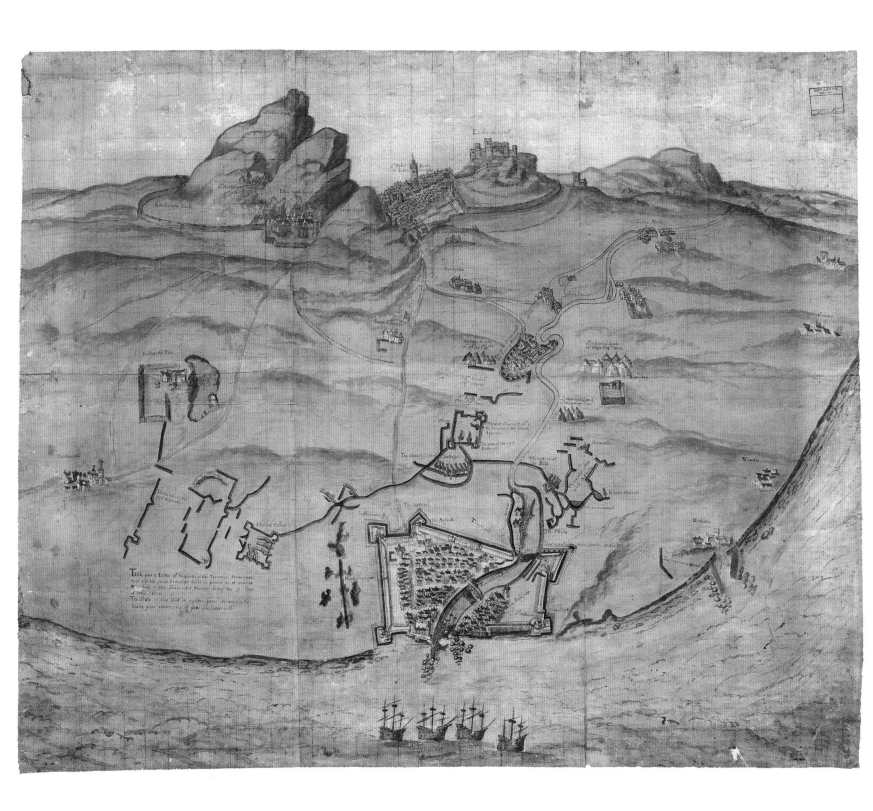

FIGURE 2.2

(*Opposite*) This stunning view from the archives of Petworth House in Sussex records the positions of mines and artillery around Leith on the day the French defenders capitulated, 7 July 1560. It is widely presumed that the plan is by Sir Richard Lee, England's chief military engineer. Lee was present during the siege and he sent 'a platt of Leith' (a bird's-eye view plan) to London on 15 May, requesting the young Queen Elizabeth's decision regarding 'works' shown on it. Lee had been in Edinburgh during the Earl of Hertford's attack on the city in 1544 and at the battle of Pinkie (1547), and had planned the *trace italienne* fortress at Eyemouth. This new style of overhead scaled plan was being introduced into Britain in the mid sixteenth century by Italians, but it took time to gain acceptance as the conventional bird's-eye view long remained the more familiar and accessible mode of presentation. Although the Treaty of Edinburgh required the dismantling of Leith's fortifications, they are shown quite clearly in the background of Greenvile Collins's chart of 1693 (fig. 2.11a), looking substantially intact.

Source: Richard Lee?, *The Plat of Lythe w' th'aproche of the Trenches Therevnto* (1560). Courtesy of West Sussex Record Office.

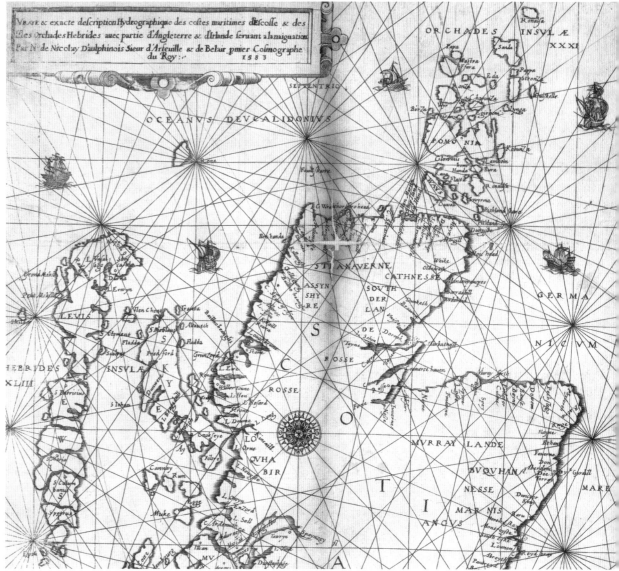

FIGURE 2.3

(*Above*) The northern part of Scotland, from Nicolas de Nicolay's map of 1583.

Source: Nicolas de Nicolay, *Vraye & exacte description Hydrographique des costes maritimes d'Escosse & des Isles Orchades . . .* (1583).

(a)

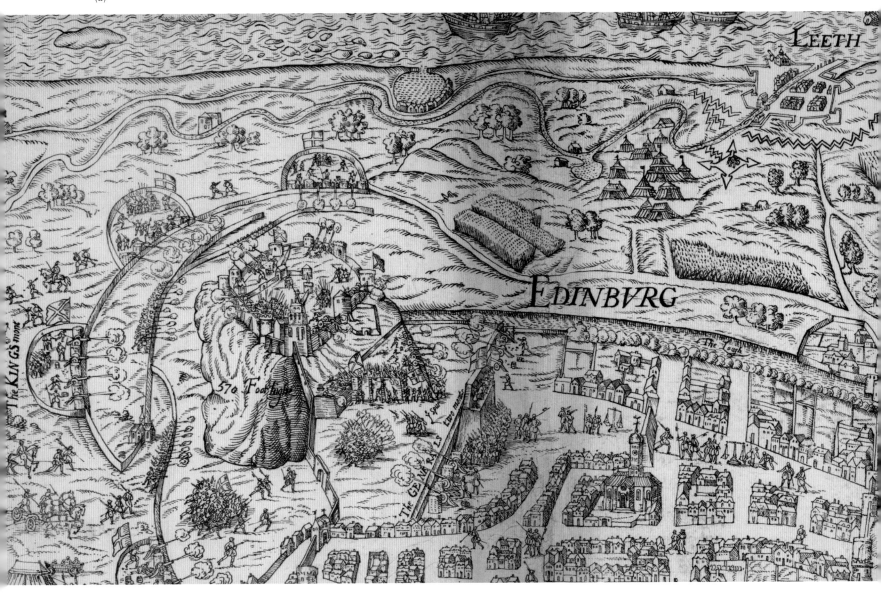

FIGURE 2.4

These two details are our best surviving depictions of the famous Edinburgh Castle spur, an innovative artillery defensive work, hastily constructed in 1548 to designs by the Italian military engineer Migiliorino Ubaldini. One of the earliest depictions of Ubaldini's 'forte on the castle hill' can be seen in the distance on Richard Lee's 'platt' of the siege of Leith (fig. 2.2), but it is shown much more clearly in these bird's-eye views of 1577 and 1647.

(a) In January 1573, Rowland Johnson, then Surveyor of Works at Berwick-upon-Tweed, drew a fine 'platt' or bird's eye view of the Lang Siege at Edinburgh Castle. This siege lasted nearly two years, part of the bitter civil war between the Queen's Men – supporters of the deposed Mary, Queen of Scots, who occupied the castle – and the King's Men, supporters of her son, the infant James VI. The siege was only broken by the arrival, in April 1573, of an English force with heavy artillery, who joined the King's Men and pounded the castle into submission. The view shows the castle at this final, dramatic stage in the siege. Although the original drawing is now lost, the copy shown here and subsequently printed in Raphael Holinshed's *Chronicles of England, Scotland and Ireland* (1577) ensured its lasting fame.

Johnson and Berwick's master gunner, John Fleming, wrote a report with an excellent description of the spur: 'Also, we fynde upon the este syde [of the castle] a Spurre lyk a bulwarke, standing befor the foot of the rocke . . . which Spurre inclosethe that syde flanked out on both sydes;

(b)

on the sowth syde is the gaite wher they enter the Castle; which Spur is lyke XX foote hye, vamyred [strengthened] with turfe and basketes, set up and furnished with ordinance.'

(b) James Gordon's outstanding bird's-eye view of Edinburgh, commissioned by the town council in 1647, shows the spur more clearly (with Johnson's besiegers helpfully removed from view). The entrance gate can clearly be seen, along with a demi-bastion at the south-west corner, crenellations and two canons ready for action. By this time, the spur had been repaired more than once. The Lang Siege saw its wholesale demolition, with rebuilding soon afterwards, along with the new Half Moon battery shown by Gordon, which replaced the original David's Tower. During the Bishops' Wars (1639–40), the Covenanters also shot at the spur and sprang a mine under its south-east angle, although it was soon repaired. The spur was finally demolished in 1650 under Parliamentary orders, and buried when the esplanade was formed in 1753. Recent archaeological excavations carried out in 2009–11 found the stonework of the spur wall in the position confirmed by these earlier plans, buried deep under the esplanade.

Sources: (a) Rowland Johnson, The siege of Edinburgh Castle (1577).
(b) James Gordon, *Edinodunensis Tabulam* . . . (1647).

September and Seymour's removal from office in October, the English rapidly withdrew from Scotland, and hostilities were formally ended with the Treaty of Boulogne in March 1550.

These dramatic events also involved new military technologies, which in turn led to an urgent requirement for maps. Historians will continue to debate the term 'Military Revolution' – the radical change in military strategy, technologies and tactics that took place in various European countries from the late fifteenth to the mid seventeenth centuries – but the central elements are not in doubt. The era saw a combination of new mobile and highly destructive field guns, which in turn encouraged new 'artillery' fortifications that could withstand them. These new artillery fortifications, often termed *trace italienne* defences after their country of origin, had thick and low walls, usually of masonry and earth, to absorb the higher-velocity cannon fire. They also had projecting angled bastions, often arrow-shaped, with two faces to house offensive gun batteries to fire into the besiegers, and two flanks, enabling gunners to cover the adjacent lengths of curtain walls and faces of the neighbouring bastions. Other components of the Military Revolution were the vastly increased size of armies and increased specialisation within them, two related developments which collectively could only be funded by more powerful, affluent states. There was also a rediscovery of some of the classical texts on military theory, which encouraged a proper drilling of infantry and better integration between the army and the navy.

Henry VIII's expeditionary forces in France quickly learned these new methods first hand, along with the need for artillery fortifications and firearms, and they brought these new practices with them to Scotland in the 1540s. The *trace italienne* fortification first appeared in the British Isles in the 1540s, and there was a very rapid implementation of gunpowder weapons, including more portable guns. New military maps were required to show these new types of artillery fortification. By the mid 1540s, Henry VIII's top mapmakers, such as John Rogers and Richard Lee, were drafting new, quite sophisticated military plans for the king to assess fortification proposals. Following Henry VIII's capture of Boulogne in September 1544, and Francis I's revenge against England, there was an influx of Italian military engineers into the English army too. Foreign chartmakers, such Jean Rotz and Nicolas de Nicolay from France, and the Scottish Highlander and spy John Elder, were also brought over to England to draw maps and gather military intelligence.

The Rough Wooing campaigns into Scotland involved the most extensive use of maps in a military context to date in Britain. Following Henry VIII's death in January 1547, Edward Seymour, by then Duke of Somerset, became Lord Protector of England. When he invaded Scotland in late 1547, he took several professional military map-makers with him, including Richard Lee, Thomas Petit and John Elder, as well as Italian military experts. These included Archangelo Arcano and Giovanni di Rosetti, known as 'Mr John the Ingineer', appointed to modernise Broughty Castle and design the nearby Balgillo Fort, an ambitious and sophisticated *trace italienne* stronghold. The Scots also made use of Italian engineers at this time, provided and financed by France. Antonio Melloni, the architect of the five-bastioned Fort d'Outreau near Boulogne, and Pietro Strozzi were both present at the siege of St Andrews in 1547. Strozzi was leader of the Italian mercenaries in Scotland, and he went on to construct the *trace italienne* fort at Leith. On 2 February 1548, Migliorino Ubaldini was appointed 'to perfect the Scots in the knowledge of arms and to organise the defence of the realm'. He was allowed 'supreme command . . . of all the Scottish forces by land and sea, and to have access to all the fortresses, munitions, etc.'. Ubaldini arrived in Dumbarton in late February 1548, and by March was in Edinburgh, where the 'Italian devisar of the forte on the castle hill' lodged for some months (fig. 2.4) before going on to work on defences at Dunbar and possibly also at Stirling (see fig. 3.5).

The battle of Pinkie

The Duke of Somerset advanced into Scotland on two fronts in 1547. The main 'Army Royal' of around 18,000 men marched from Berwick (supported by an armada of eighty ships), while in the west, 2,000 foot and 500 horse under Lord Wharton and the Earl of Lennox took Annan and established garrisons in Castlemilk and Moffat (fig. 2.5). The battle of Pinkie Cleugh was the last pitched battle fought between the Scottish and English armies, and took place on the banks of the River Esk south of Musselburgh on 10 September 1547 (fig. 2.6). The new elements of the army, including 6,000 heavy cavalry and German, Spanish and Italian armed infantrymen, proved to be of vital importance to the English victory.

Advancing from the west towards Pinkie, the Earl of Arran deployed the Scottish army on earthwork fortifications defending the line of the River Esk. Estimates of the army's size range from 20,000 to 36,000, and although numerically superior, it mainly consisted of traditional archers and pikemen (armed with long, thrusting spears). In military terms, an English Renaissance army with professional soldiers and firearms met a Scottish medieval one, armed with pikes, bows and arrows. The most mobile elements were on the wings – the Argyll Highlanders to the north and the Border light horsemen to the south, who had been badly mauled in a cavalry skirmish the day before.

Somerset knew that controlling Carberry Hill and St Michael's Church in Inveresk could give vital vantage points for his artillery, but in moving towards these positions, he was surprised by Arran taking the offensive, hoping that his pikemen could make a swift blow against the English flank. Arran's plan did not work, and the pikemen were repulsed by heavy guns from the English ships in the Forth and a cavalry advance. In spite of heavy English cavalry losses, the Scots advance was stalled, and an attempt to regroup failed, resulting in a retreat, which rapidly become a rout. More Scots died in the rout than in the battle itself, as they were pursued for five miles towards Edinburgh, with some 6,000 killed and perhaps 1,500 taken prisoner. The English eyewitness William Patten almost seemed to revel in the horror of the carnage, with 'dead corpses lying dispersed abroad, some their legs off, some but houghed, and left lying half-dead, some thrust quite through the body, others the arms cut off, diverse their necks half asunder, many their heads cloven, of sundry the brains pasht out, some others again their heads quite off, with other many kinds of killing'. Yet whilst this was a dreadful defeat for the Scots, Somerset had failed in his central ambition of the marriage proposal, and effectively won only the chance to plant temporary garrisons in southern Scotland.

FIGURE 2.5

(*Overleaf*) This manuscript map or platt records the English 'pale' in south-west Scotland, the West Marches, created between September and November 1547, and the main strongholds occupied by the English. The platt looks towards the Solway in the south-east ('the See'), and is centred on Castlemilk, which was taken on 10 September 1547 by an English force under Lord Wharton, Warden of the West Marches. The next day, the Scottish forces in Annan steeple submitted, and the town (directly above Castlemilk on the map) was badly damaged – 'anand towin destroyit'. By 17 October, Dumfries (on the right-hand edge) had been taken, and the tower of Cockpool (between Castlemilk and Caerlaverock to the upper right) surrendered into English hands on 19 November.

We know that, on 26 October, Wharton requested from Edward Seymour, Lord Protector Somerset, that a surveyor be sent to make a platt of the pale of Lochwood, and to assist with intelligence to support an assault on Kirkcudbright. Two engineers, William Ridgeway and Thomas Petit, are recorded in the area after 12 November. It seems probable that this platt is by Petit, on stylistic grounds (even though another hand wrote the names), due to its similarities to a map of Calais by Petit (1646–47).

Petit was appointed surveyor of Calais from January 1546, a post he held well into the 1550s, but worked on several other Scottish castles, including Wark (1543), Ross in Bute (1544), Burntisland (1548) and Haddington (1548). The platt was presented to Somerset in December 1547, and after Somerset's execution in 1552, kept firmly in the possession of William Cecil, Lord Burghley.

The platt is a valuable record of the specific sites occupied by the English, from Carlisle (upper left) and Langholm (lower left) across Annandale and into Dumfriesshire. The 'debatabill landis' are recorded by the River Esk on the far left side, between the 'lyddisdale armestrounges' and the 'grahamis upoun esk' on the other side of the river. The platt gives clear bird's-eye details of the strongholds, with descriptive notes, and the distances between them in miles recorded as rays. Many of these castles and tower houses were strengthened and refortified at this time, and used as bases for harrying the surrounding countryside.

Source: Thomas Petit?, The platte of Milkcastle (1547). Courtesy of Hatfield House.

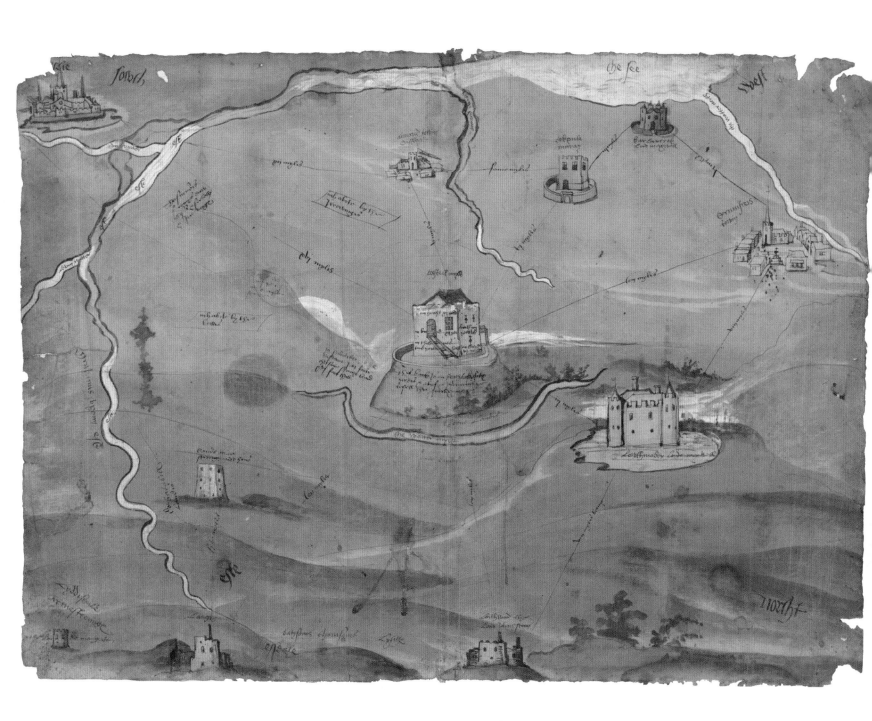

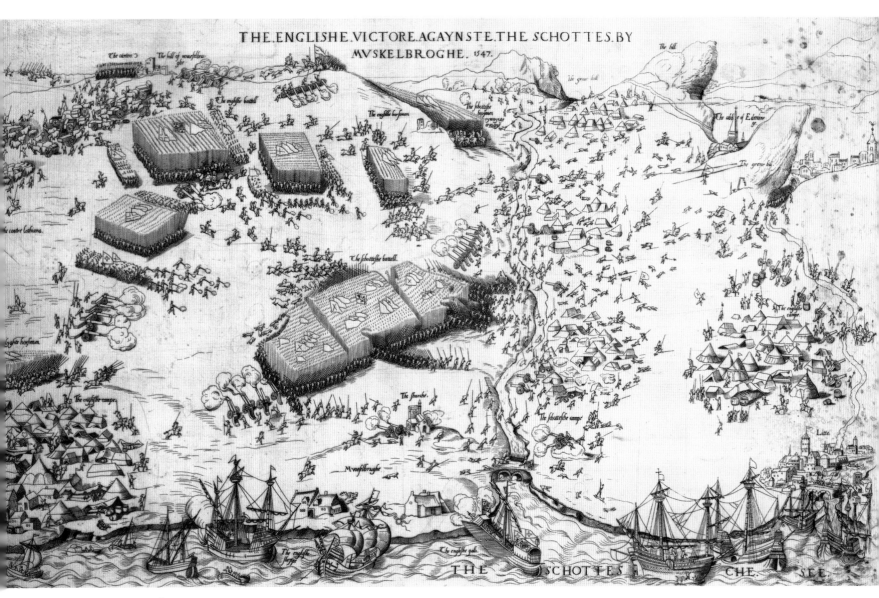

FIGURE 2.6

These three very different views or maps of Pinkie were all made within a year of the battle. They were intended for English audiences, part of the wider propaganda war of the Rough Wooing, which included publications justifying the English control of Scotland, as well as the merits of the Protestant religion. The three maps offer complementary insights into the battle, as well as illustrating three different production methods.

(a) (*Above*) The copperplate engraving, attributed to Thomas Geminus alias Lambrechts, is a very early example of a news map, and combines several stages of the battle into one.

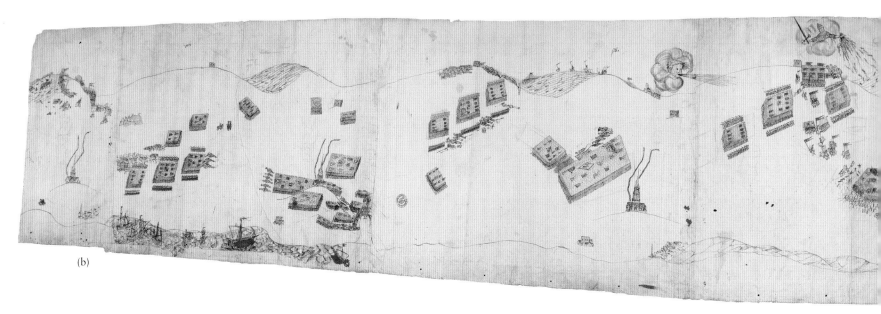

(b)

FIGURE 2.6 (*continued*)

(b) (*Above*) The painting by 'John Ramsay Gentyllman, without money' on five panels, showing the consecutive phases of the battle, is thought to be a copy, or perhaps an artist's preparatory sketch, for a commemorative painting honouring the Duke of Somerset. Ramsay served with Somerset, and the five scenes show the following:

1. The cavalry skirmish of 9 September, with Inveresk Church in the foreground.
2. The English army in order of battle, with Lord Protector Somerset to the top left and supported by the fleet in the Forth, moves towards the Scottish army, passing Inveresk Church in three divisions. Pinkie House is in the foreground, with Carberry Tower beyond Inveresk Church.
3. The Scottish army mass together, attacked by the English cavalry charge.
4. The English cavalry charge is repulsed, but the fire of the guns and archers begins to break up the Scottish column.
5. The rout of the Scots, fleeing towards Arthur's Seat and Edinburgh.

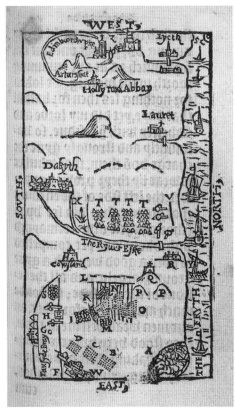

(c)

(c) (*Left*) William Patten's woodcut accompanied his detailed and entertaining account of the battle, which was published to high acclaim and interest in 1548. Although seemingly crude at first glance, it gives an unusually early overhead 'plan' view of the battle, illustrating most of the central elements and places: A the English camp; B, C, D the English foot; G Somerset; I the English horse attacking Angus; L the lane; M the Scottish forward and horsemen; N the Scottish battle; O the Scottish rearward; PP two hillocks; Q Inveresk Church; R Musselburgh; TTTT the Scottish tents; V turf wall; W the English baggage; X marsh.

Sources: (a) Thomas Geminus alias Lambrechts, *The Englishe Victore agaynste the Schottes by Muskelbroghe* (1547). (b) John Ramsay, Battle of Pinkie commemorative scroll (*c.* 1548?). Courtesy of the Bodleian Library, University of Oxford. (c) William Patten, 'Battle of Pinkie', from *The Expedicion into Scotla[n]de of the most Woorthely Fortunate Prince Edward, Duke of Soomerset* (1548).

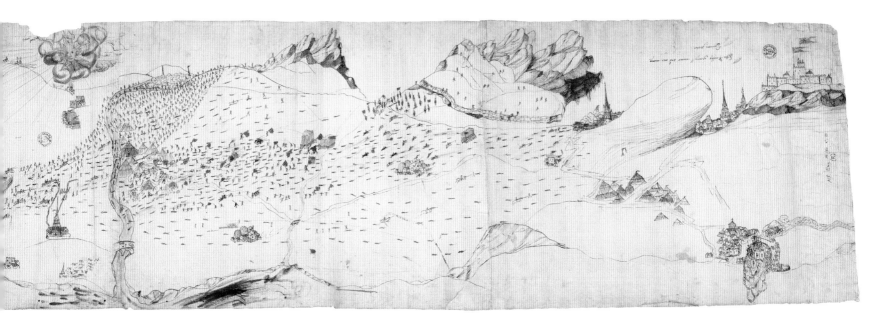

The siege of Leith and the Debatable Land

Leith's strategic importance as Edinburgh's port made it a focus for English attempts to occupy and defend it in 1547, after the battle of Pinkie, and test trenches were dug but not developed further. The Scots looked to the French for assistance, and from June 1548 French troops arrived in Leith, soon numbering over 8,000 men. The Italian military engineer Migiliorino Ubaldini, who also designed the spur at Edinburgh Castle, planned a bulwark at the Leith Kirkgate and chapel near the harbour around this time. But it was left to another Italian engineer, Pietro Strozzi, to continue his work. Over the next decade, Leith became the first town in the British Isles to be wholly enclosed by a *trace italienne* style of fortification, with low, thick walls designed to absorb rather than resist cannon fire, and with eight large, projecting bastions providing covering fire for one another (fig. 2.2).

The Siege of Leith (1560) finally brought to an end a twelve-year encampment of French troops at Leith, and with it the broader French military occupation of Scotland from 1548. But the excellent fortifications made it a tough job. Two temporary 'fortlets', with four corner bastions, were also constructed (Mount Pelham, to the left or east of Leith, and Mount Somerset, above, to the south) to enable artillery to fire into the town. On the morning of 7 May there were two breaches made in the walls, but the scaling ladders proved too short and the English were defeated, with several hundred Scots and English killed. Mines were then dug towards the fortifications, clearly shown in the plan. In the end, however, the siege was lifted through peace talks culminating in the Treaty of Edinburgh, which secured the withdrawal of French *and* English troops from Scotland; by 17 July, the foreign soldiers had all departed.

During their time in Scotland in the 1550s, the French were also involved in settling scores on the Anglo–Scottish border. About ten miles north of Carlisle, running inland between the River Sark and River Esk, lay the 'Debatable Land', a strip of country around four miles wide by twelve long. It was described as such from at least the fifteenth century, as its ownership was disputed between Scotland and England. As a result, neither side had complete dominion, and it became a refuge for some of the worst elements of the frontier – a wild, lawless and dangerous place. But it was excellent farmland, and attempts to devastate the area, to render it unfit for habitation, always had a temporary result as people soon came back. The growing problems of the area were eventually dealt with when the Rough Wooing campaigns were finally over. England proposed an easy solution of claiming the whole area, but the Scots countered with a proposal 'so that ilk realme might ken their awin part and puniss the inhabitants thereof' provided that Canonbie should be entirely in Scotland. The Treaty of Norham (concluded on 10 June 1551) agreed that the land should be divided between the two countries by means of a special boundary commission, one of the first in European history. Following much wrangling, the arbitration was finally concluded on 16 August 1552, adopting the French suggestion (fig. 2.7).

Nationwide reconnaissance and topography

We do not know the exact motivations behind the monumental surveys of Scotland undertaken by Timothy Pont some time between his graduation from St Andrews in 1583 and his death around 1614. There is no doubt, however, that Pont's maps would have been useful politically for king and state; James VI's attempts to control unruly Highland clans and to deal with widespread lawlessness along the Anglo–Scottish border would have been greatly facilitated by detailed maps. Within sixteenth-century Europe, maps were becoming essential tools for statesmen, too. In England at this time, William Cecil, First Baron Burghley and chief adviser to Queen Elizabeth throughout most of her reign, collected and made regular use of maps. Several detailed Scottish military plans (for example, fig. 2.5), as well as regional maps of the Borders at similar scales to Pont's work, only survive today through their preservation in Cecil's collection, now at Hatfield House.

Pont's maps did not have a primary military function behind them, and we can see that the breadth and range of their content had a much broader purpose, describing regions and the lie of the land. That said, in being the first detailed systematic recon-

FIGURE 2.7

(a) (*Opposite*) On 24 September 1551, Henry Bullock, a practised land surveyor, executed this site map for dividing the Debatable Land. It is an early example of an international boundary dispute map, as well as being an early example of a scale map in a British context, reflecting the importance of showing the divided lands fairly. It clearly shows the rival lines proposed to join the Esk and the Sark: to the north is the line marked as the 'English Commissioner's Offer', and to the south is 'the Scottes offer'. The map also shows the final resolved line, indicated on the map as 'accord with the French ambassador'. Bullock was paid a sum of 20 nobles (around £1,500 today) and went on to become Master Mason of the King's Works.

(b) (*Below*) By the spring of 1553, the Scottish government had gathered a work crew 'to caus the dikkes and fowseis of the Debatable lande be biggit'. It was a significant landscape feature – a boundary wall and mound with distinct markers – recorded on maps through to the present day. Blaeu's engraved maps, with their simpler and standardised set of symbols for representing the ordered landscape, tended to omit earthworks or mounds in the ground, but even they include the 'March Dyik'.

Sources: (a) Henry Bullock, Map of the West Border Land (1552). Courtesy of The National Archives. (b) Timothy Pont / Joan Blaeu, *Lidalia vel Lidisdalia regio*, Lidisdail (1654).

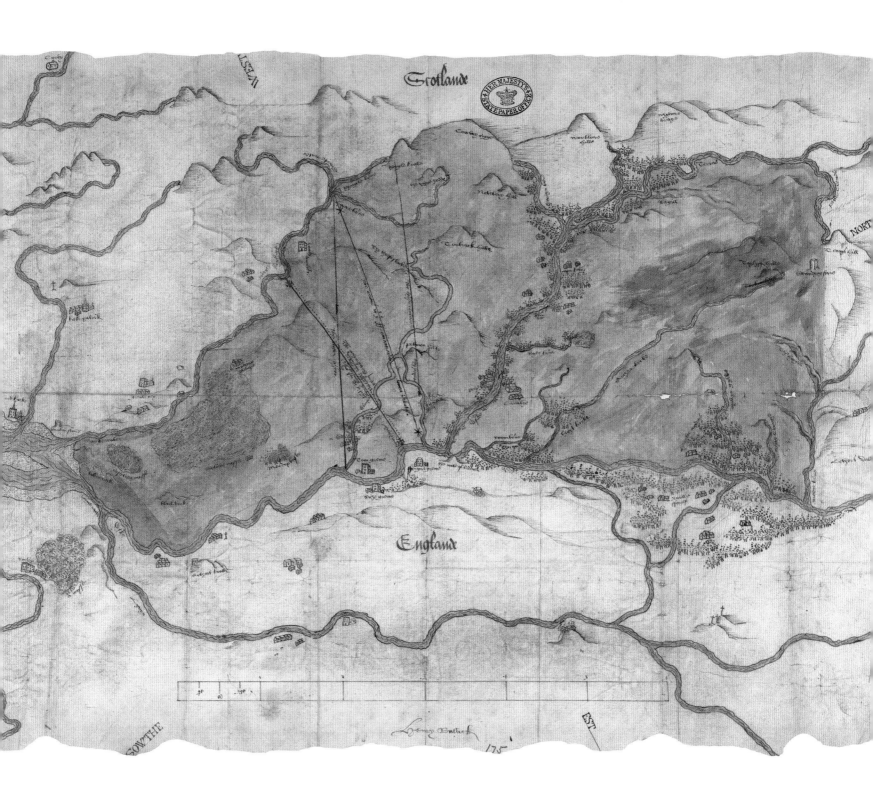

naissance of Scotland, they provide excellent detail of many castles, tower houses, forts and other defensive structures (fig. 2.8). While many of these fortified residences were not being actively attacked or defended during Pont's time, they had been recently, and were to see further military action the following century. Getting Pont's work into print, through the Blaeu *Atlas Novus* of 1654, was also significantly disrupted and delayed by war – the Wars of the Three Kingdoms, including the Bishops' Wars of 1639–40, the Scottish Civil War of 1644–45, the campaigns of James Graham, 1st Marquis of Montrose (fig. 2.9), as well as the Anglo–Dutch Wars of 1652–54. John, Earl of Lauderdale, had promised a description of Lauderdale but was captured at the battle of Worcester (1651). Blaeu solemnly records this, apologising for the missing description with a note to 'Enjoy these, Reader, until he has been restored, or some other has provided better.'

Cromwell and the Protectorate citadels

Oliver Cromwell's invasion of Scotland, with a crushing defeat for the Scots at the battle of Dunbar (1650) and a general sack of Dundee (1651), began an eight-year period of military occupation (fig. 2.10). A set of massive and imposing citadels were constructed at four selected locations: Ayr, Inverness, Perth and Leith, backed up by a system of garrisoned forts, widely distributed throughout the countryside. Although, sadly, no contemporary drawings survive of the citadels, which were badly damaged following the Restoration in 1660, they are often shown on later maps, allowing a reasonably good understanding of them (fig. 2.11 and 2.12). The citadels were quite unique in style and size, as well as being expensive, and it is possible none were ever fully completed. All were regular in plan, pentagonal in shape and monumental in scale, with internal elevated central buildings, officers' houses and barracks, and a chapel. The smaller garrisons occupied many traditional state castles, including Edinburgh, Stirling and Tantallon, as well as smaller castles such as Blair and Dunstaffnage, and included infantry and cavalry units. Some sketch plans survive, showing plans for construction or reconstruction of these smaller garrisons, such as at Stornoway/Steòrnabhagh and Duart (fig. 2.12). The garrisons and forts were an understandable solution to the problem of attempting to control an alien and resentful people, but they did not last long. With the death of Cromwell in 1658, the Commonwealth fell into a period of instability, which ended in 1660 when the English army occupying Scotland marched south under the command of General George Monck, inviting Charles II to return as king.

Emergence of a military engineering institution in Scotland

By the late seventeenth century, deliberate attempts were made to establish military engineering as an institution in Scotland, along the lines of the English Board of Ordnance, with emerging responsibilities for military draughtsmanship (see chapter 3). This saw the establishment of the post of Chief Engineer in Scotland from 1671, with the appointment of John Slezer. Slezer was a talented military draughtsman, who probably came from a German-speaking part of Europe, and who first visited Scotland in 1669. Slezer's contact with the nobility, including the earls of Argyll and Kincardine, as well as his clear expertise as a military draughtsman, encouraged expanding the duties of the fledgling post. He was given responsibilities for supervising the Ordnance (military infrastructure, defences and equipment) in 1677, and in the following year he drew up a plan 'to make a book of the figures, and draughts, and frontispiece . . . of all the King's Castles, Pallaces, towns, and other notable places in the kingdom', a work eventually published in 1693 as his *Theatrum Scotiae*. It is clear that the work grew out of Slezer's surveying of military sites, such as Dunnottar (fig. 2.13), and the *Theatrum* was intended to supplement his income by including civilian and antiquarian sites of interest. Sales of the *Theatrum* sadly failed to cover costs, and Slezer ended his days in the debtor's sanctuary in Holyrood Abbey. However, he was able to keep his military duties until 1716, when the army was reorganised.

FIGURE 2.8

As shown by Timothy Pont, Duffus Castle and Spynie Palace were both impressively powerful fortified residences, standing on the edge of the Loch of Spynie, which was once a sea loch with a good safe anchorage. The loch silted up over time and was drained in the early nineteenth century, so Pont's map gives a useful impression of its much larger extent in the sixteenth century.

Duffus Castle was a motte-and-bailey castle, dating from the twelfth century, one of the strongest castles in Scotland, and a fortified residence for more than 500 years, from the 1100s to the 1700s. The central stone keep shown by Pont was constructed in the late thirteenth or early fourteenth century, when the castle passed by marriage to the Sutherland Lord Duffus, whose family held it until 1843. It was attacked several times in its long history, initially in 1297 by the Scots, in 1452 by the Douglas Earl of Moray and in 1645 by Royalists.

Spynie consists of a massive fifteenth-century keep within a large courtyard. As clearly detailed by Pont, the central keep, David's or Davie's Tower, rises six storeys with a garret at the top. The enclosing courtyard walls are also flanked by square towers. It was the fortified residence of the Bishops of Moray, and probably constructed by Bishop Innes in the late fifteenth century. During Pont's time, the castle and estate were in the hands of Alexander Lindsay, 1st Lord Spynie, and additional fortifications were added as protection against a perceived threat from the Spanish.

After the Reformation, Spynie's owner Bishop Guthrie was deposed in 1638, but he tried to keep possession of the palace, which he had provisioned and garrisoned. However, it was attacked by General Munro in 1640, and Guthrie was eventually forced to surrender. Gordon of Huntly unsuccessfully besieged the palace when it was held by Covenanters in 1645, while acting for the Marquis of Montrose. Following the restoration of the Episcopy to the Scottish Church in 1662, ownership of the castle passed back to the Church, but the building by then was starting to fall into decay.

Source: Timothy Pont, Detail of Duffus Castle and Spynie Palace, [Gordon] 23 (*c.*1583–1614).

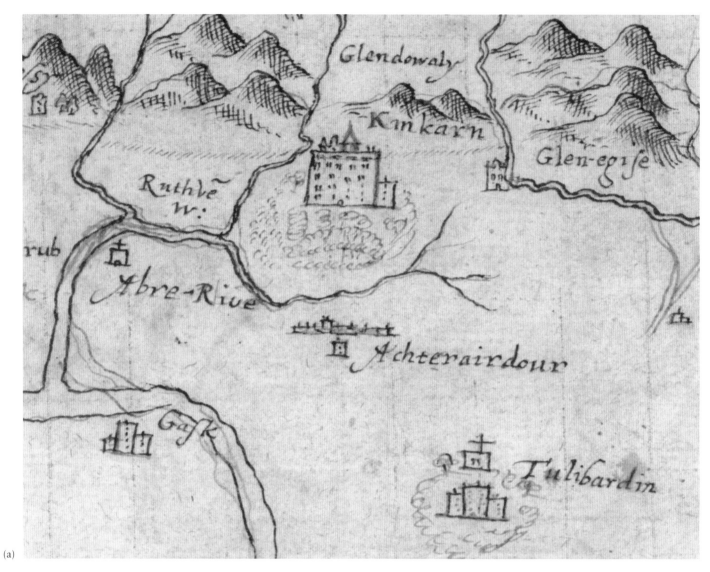

(a)

FIGURE 2.9

(a) This detail from Timothy Pont's map of Strathearn is a uniquely important depiction of the impressive Kincardine Castle, seat of the Grahams of Montrose from the mid thirteenth century. Pont's manuscript map is orientated with south at the top, depicting the castle around a mile south-east of Auchterarder. As the main seat of James Graham, 1st Marquis of Montrose (1612–50), the castle was largely demolished by the Campbell Earl of Argyll in 1646, following Montrose's defeat at Philiphaugh.

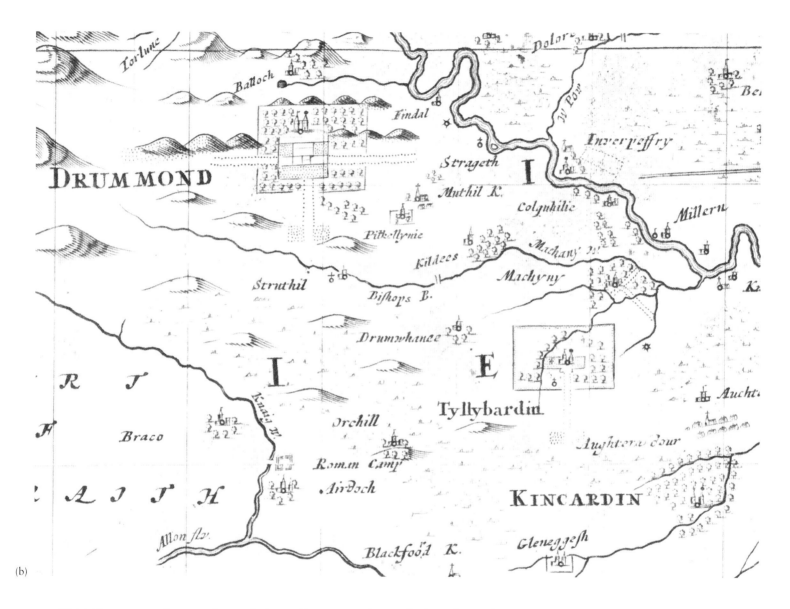

(b) Adair's map surveyed a century later, and orientated in contrast with north at the top, shows the major houses of Drummond and Tullibardine, with enclosed grounds around them, but not for Kincardine.

Although Montrose initially joined the Covenanters in the Wars of the Three Kingdoms, he subsequently supported King Charles I as the English Civil War developed. From 1644, he was appointed Lord Lieutenant of Scotland, and he went on to fight two highly successful campaigns, distinguished by rapidity of movement, defeating his opponents in six battles. Following three years of exile in Norway, he was captured in Assynt and brought to Edinburgh, where he was executed. His head was removed and stood on the 'prick on the highest stone' of the Old Tolbooth outside St Giles Cathedral from 1650 until the beginning of 1661.

Sources: (a) Timothy Pont, Detail of Kincardine Castle, Strathearn, Pont 22 (c.1583–1614). (b) John Adair, *The Mapp of Straithern, Stormount, and Cars of Gourie, with the Rivers Tay and Iern* (c.1720).

FIGURE 2.10

This map by Theodore Dury, orientated with south-west to the top, dates from around 1690 and records the contemporary state of Blackness. Dury's plan also clearly shows how the earlier medieval castle dating from the fifteenth century was adapted and refortified in the sixteenth century in response to the growing threat of gunpowder artillery. Blackness was originally built for the Crichton family as a lordly residence in the 1450s. When Sir James Hamilton of Finnart became Master of Works in 1530, he considerably strengthened the external walls and entrance gate. The visibly massive external walls were refortified from 1.5 to 5.5 metres thick, and a caponier (a low-level gallery with gunholes, 'D') and spur were constructed around the entrance ('A') on the east side. The French were evidently impressed, using the castle in the 1540s as an ammunition depot, while they based their main garrison at Leith (fig. 2.2). A century later, as a royal castle, Blackness was bombarded from land and sea in March 1651 by Cromwell's forces. Most of the damage was done from a battery to the south, recorded as 'a Redoubte, built by the English, when they werre Masters of Scotland'. The garrison surrendered within a day and the castle was left in ruins, before further rebuilding after the restoration of Charles II in 1660. The new barrack proposed by Dury ('T' on the map) was not built. Instead, the central tower, called the Main Mast ('N'), which had been a prison, became a barracks from 1707. As Dury notes 'In the Tower called the Main Mast are four Chambers one above another without any Closets.' Money was clearly not to be wasted on barrack accommodation luxuries.

Source: Theodore Dury, *Plan of the Castle of Blackness* (c.1690).

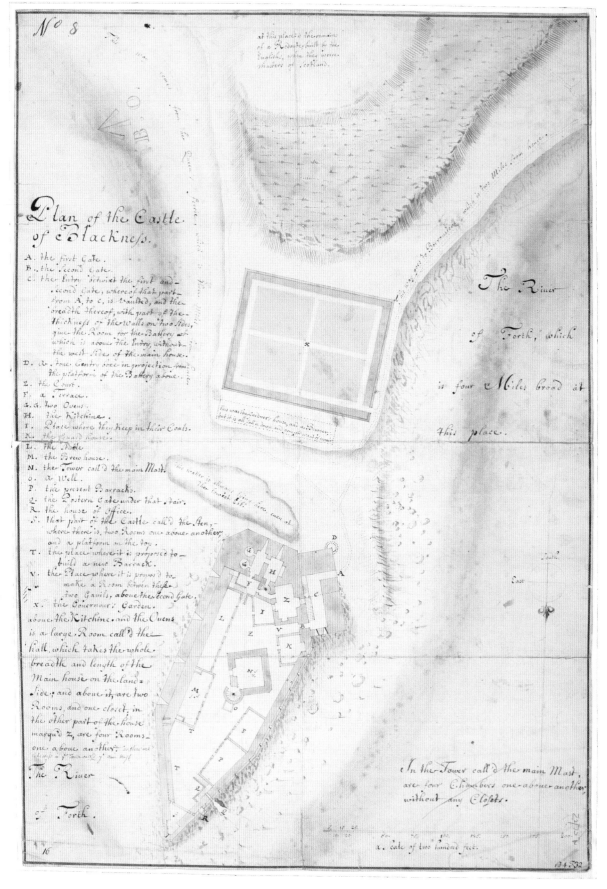

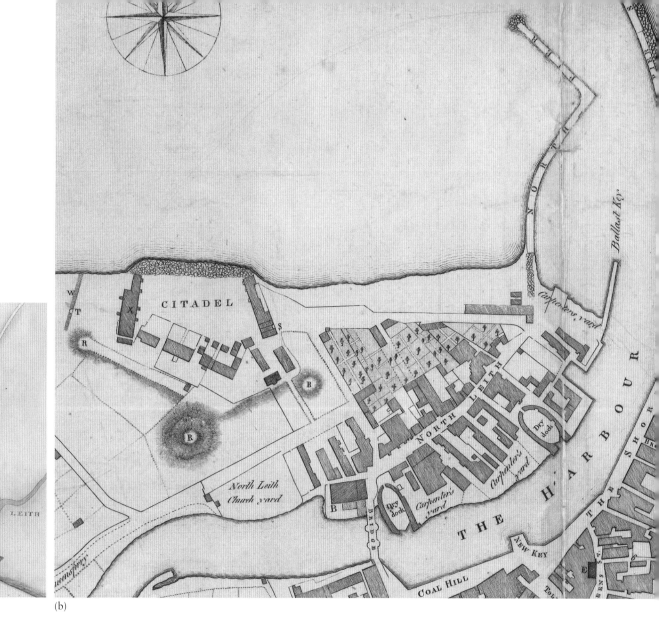

(a) (b)

FIGURE 2.11

Leith Citadel was built on Cromwell's orders from 1656 under General Monck 'to keepe in awe the chief city of this nation'. In contrast to the other citadels (Ayr, Inverness and Perth), which were built on virgin ground, several buildings were swept away for its construction, including the former parish kirk of St Nicholas. John Mylne, Master Mason to the Crown, worked on the citadel before going on to work on the citadel in Lerwick (fig. 5.17). The citadels, as monuments to the Cromwellian interregnum, did not survive long after the Restoration, but their massive size resulted in more of a lingering reduction and plunder of stone over time.

(a) Only three decades later, Greenvile Collins's chart of Leith from 1693 shows most of the north side of the citadel already washed away, through coastal erosion, and the citadel rather outshone by the remains of the French fortifications of the 1550s.

(b) Less than half the citadel was left a century later on Alexander Wood's plan of 1777, but Wood provides other topographic details of interest – 'S' and 'T' respectively mark the eastern and western sides of the citadel, and in between these, 'X' marks 'Houses said to have been Barracks for Cromwell's soldiers'. On the remaining three corners of the pentagon, 'R' marks 'Mounts raised by Oliver Cromwell when he fortified the Citadel'.

Sources: (a) Greenvile Collins, Map of Leith from the North (1693).
(b) Alexander Wood, To the Magistrates, the Commissioners of Police and the Four Incorporations, this Plan of the Town of Leith from an Actual Survey (1777).

FIGURE 2.12

These two rapidly drawn sketch maps show plans for Cromwellian defensive works in 1653, refortifying Castle Duart in Mull on the left, and constructing a new fort in Stornoway/Steòrnabhagh on Lewis/Leòdhas on the right.

The MacLeans of Duart supported the Royalists, and Sir Hector MacLean, the 18th Chief, was killed at the battle of Inverkeithing in 1651, which was a decisive victory for Cromwell's forces. Castle Duart was subsequently besieged in 1653, when Cromwell sent a task-force of six warships to the castle. Although the family had fled, a great storm hit the warships soon after they arrived at Duart, and three ships were wrecked. The map dates from this time, showing an outline of the existing castle at the top, with the prominent addition of a half-moon battery on the lower side, noted as 'd. a halfe moone ordered now to be built to lay ordinance on, & to be 33 foot wide 21 foot high'. There is no evidence that this was constructed. The great storm resulted in serious losses of men and provisions, and the surviving ships were sent to London for repairs, so it seems likely that this curtailed these construction plans. When Duart was taken over by Clan Campbell in the 1670s, it was subsequently used in a similar way, as a strategic outpost, and plans of the castle in the early eighteenth century just show the original rectangular defences (see fig. 1.10).

Stornoway/Steòrnabhagh might have been spared by Cromwell were it not for the rash actions of the young Earl of Seaforth, Kenneth Mòr Mackenzie, who in May 1653, at the age of 18, seized a privateer, the *Fortune*, in Stornoway/Steòrnabhagh harbour in an attempt to win the crew over to the Royalist cause. The reprisals that followed saw warships and provision ships under the command of Colonel Cobbet sail from Leith to take possession of Lewis/Leòdhas and construct this garrison fort. The fort occupied the whole of the Stornoway/Steòrnabhagh peninsula, including the former church of St Lennan (b) which became a dormitory for soldiers. The sketch also shows a manor house probably connected to the earlier fisheries (a),

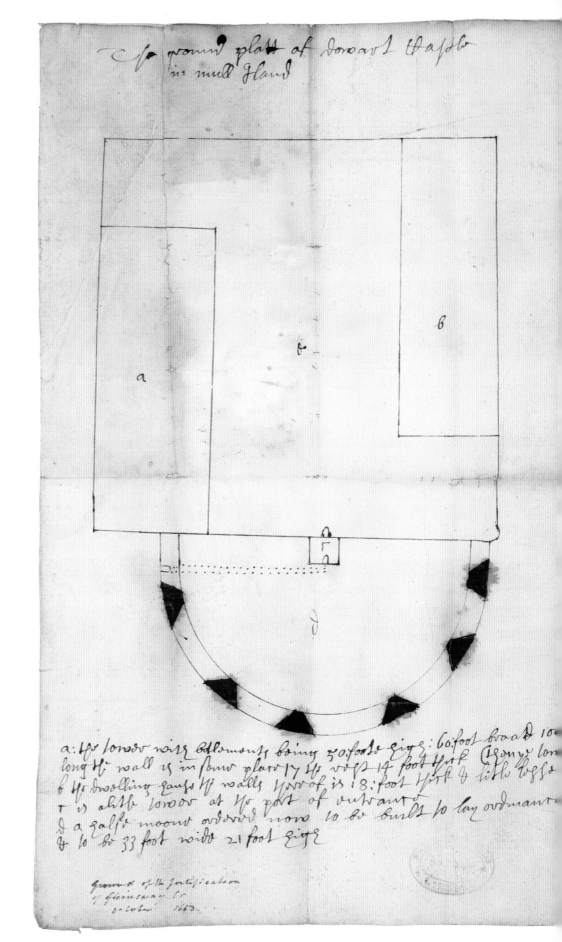

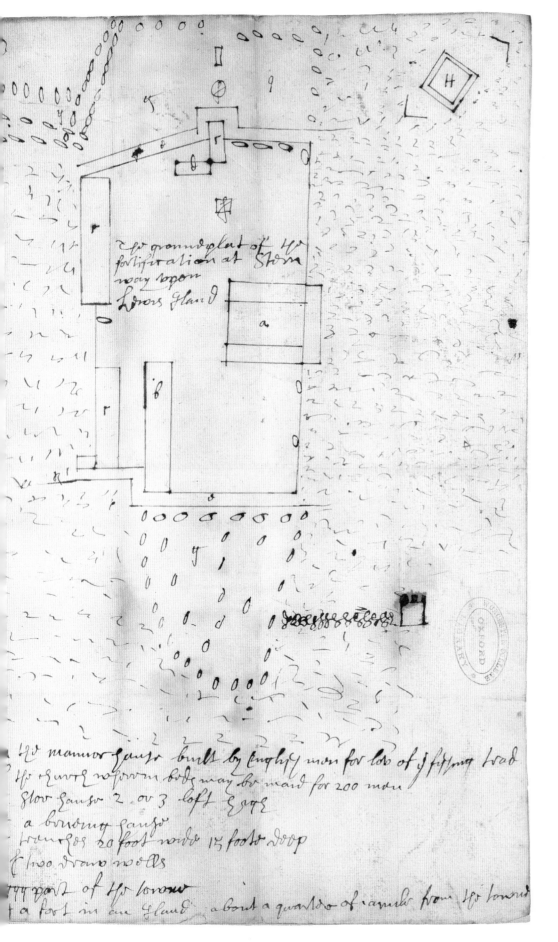

storehouses, two or three storeys high (c), a brewing-house (d) and a couple of draw wells (f). The north and south sides of the fort were protected by water or slimy mud, and the lines of the trenches (e), approximately on the lines of the current Cromwell Street and James Street, protected the west and east sides. Lack of suitable timber led to a plunder of stone from existing buildings, and many of the soldiers in the four companies of the garrison lived in the town (g). The diamond-shaped feature (h) records the Eilean na Gothail/Goat Island fort, constructed by Mackenzie using armaments from the old castle.

a. The manor-house built by Englishmen for love of the fishing trade
b. The church wherein beds may be made for 200 men
c. Store-houses two or three loft (storeys) high
d. A brewing-house
e. Trenches 20 feet wide, 15 feet deep
f. Two draw-wells
g. Part of the town
h. A fort in an island about a quarter of a mile from the town.

These two maps of Duart and Stornoway/Steòrnabhagh appear together as they were both visited by same Cromwellian task force. The fleet landed in Stornoway/Steòrnabhagh in late August 1653, before sailing on to seize Dunvegan Castle in Skye. They arrived at Duart by 5 September, only to be struck by the storm on 22 September. The maps survive through the collections of Sir William Clarke (1623?–66), Secretary of War to the Commonwealth and Charles II, whose son, George Clarke, donated it along with his other collections to Worcester College in Oxford, where the map survives today.

Source: Anon., *The ground platt of Dowart Castle in mull Island. The ground platt of the fortification at Stornoway upon Lewis Island* (1653). Courtesy of the Provost and Fellows of Worcester College, Oxford.

Plan of Dunotter

Inner Court

Carlihof

Bowlinggreen

A is ye Entrie
B ye Tiddel
C ye Tower
D ye Ammunition
  houffe

A: 1675

In October 1688, Slezer, as captain of the Scottish artillery train, was ordered to proceed to England with the rest of the Scottish forces, against supporters of William of Orange, who became King of England the following year. The so-called 'Glorious Revolution' which deposed King James II of England (or VII of Scotland) was a time of rapidly changing loyalties in politics and religion, and following the Jacobite victory at Killiecrankie in July 1689, Slezer was forced to take an oath of fidelity to William. This was accepted by the new Williamite regime in 1690, and he was then commissioned as 'captain of the Artillery Company and surveyor of Magazines'. Military forces in Scotland increasingly now came under the control of the British army, a process formalised by the Act of Union in 1707, which transferred responsibility for Scottish defences to the British Board of Ordnance. The Jacobite supporters of the deposed King James and the exiled House of Stuart went on to launch a series of rebellions and uprisings that would dominate the next half century, and Scotland became a new focus of military activity to counter this. These related developments brought with them a new era of military mapping in Scotland.

FIGURE 2.13

(*Opposite*) This little-known and unique hand-drawn plan and view of Dunnottar Castle by John Slezer dates from his early years in Scotland (1675). Slezer was appointed Chief Engineer in Scotland from September 1671, but only appointed lieutenant of the artillery from March 1677, with its more practical responsibilities for supervising the Ordnance. The plan is a clever juxtaposition of the bird's-eye perspective of the surrounding coastline, with an overhead plan of the castle buildings themselves, and is signed 'By thir Ma[jestie]s Cheef Ingeneer for the Kingdomme of Scotland, J. Slezer'.

Dunnottar occupies a spectacular, and almost impregnable, cliff-top location, about two miles south of Stonehaven, and was no doubt fortified from early times. The Marquis of Montrose had unsuccessfully besieged it in 1645. Following a visit from King Charles II in 1650, and Cromwell's invasion of Scotland in 1651, Dunnottar was considered a good place to store the Scottish crown jewels. A decade after Slezer's drawing, the castle was used as a prison for Covenanters, some of whom were tortured and subsequently died. In 1689, when the castle was held for William and Mary, many Jacobites were imprisoned there. Following its owner the Earl Marischal's support for the Jacobites in 1715, it was partially destroyed in 1716 and 1718.

The plan clearly picks out the essential buildings and their layout. The entrance to the castle (A) at the bottom (to the west) is overlooked by outworks (B) 'ye Fiddel' (the Fiddle Head promontory). The entrance passage then turns sharply to the left, running underground through two tunnels to emerge near the fourteenth-century tower house (C) and ammunition store (D). Further east, towards the top, are the main Palace Buildings, dating from the later sixteenth century, with three quadrangular wings arranged around an Inner Court[yard] and well, and overlooking the bowling green 'Boulingreen' and 'Oartchyerd'.

Source: John Slezer, *Plan of Dunotter* (1675).

FIGURE 2.14

(*Overleaf*) John Ramsay, Battle of Pinkie commemorative scroll (c.1548?). Courtesy of the Bodleian Library, University of Oxford.

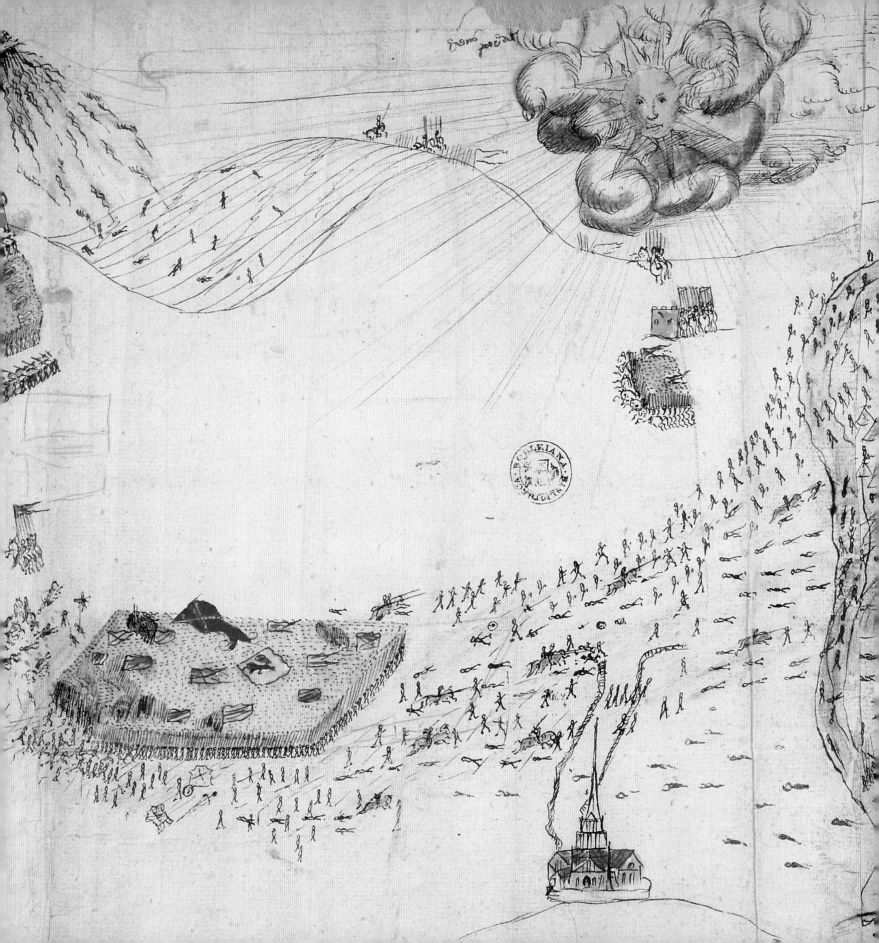

CHAPTER THREE
KILLIECRANKIE TO THE BATTLE OF GLENSHIEL, 1689–1719

On 13 April 1689, four days after the joint succession of William of Orange and Mary Stuart to the throne of Britain, John Graham of Claverhouse, Viscount Dundee, raised the Scottish Royal Standard on Dundee Law in support of the exiled King James VII. Dundee's actions initiated the first Jacobite rising in Scotland. Despite an unexpected victory for Dundee's forces at Killiecrankie on 27 July (fig. 3.1), where Dundee himself perished, the rebel army's progress was checked at the battle of Dunkeld shortly afterwards. Nevertheless, following this initial defeat of the English army, Jacobitism took hold among the clans and, thereafter, a bitter war of raid and counter-raid ensued in and around the Highlands.

Measures to suppress Jacobitism were invoked by Hanoverian monarchs and their governments over the next sixty years, but with mixed success. Government forces were increased in the old fortresses of Scotland, at Edinburgh, Stirling and Dumbarton Castles, and a new fort – Fort William (fig. 3.2) – was built at the southern end of the Great Glen. Concerns about the condition of Scotland's defences prompted orders for engineers to make plans and estimates of what needed to be done to secure 'North Britain'. But successive Jacobite risings provoked a military expansion into the Highlands that saw not only the design and construction of more forts, but also methods to improve the mobility of the British army. With

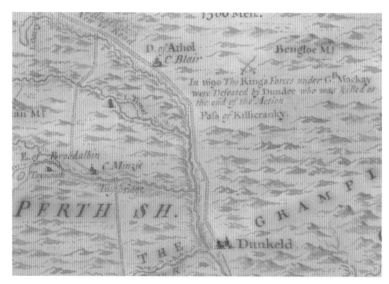

FIGURE 3.1

Although there are no surviving contemporary battle plans of Killiecrankie, its name and location are clearly shown on several Board of Ordnance maps of the early eighteenth century. Clement Lemprière may have noted the wrong year, but the 'Pass of Killicranky' where the battle took place on 27 July 1689 is correctly marked. The 'new' military road from Dunkeld to Dalnacardoch, shown here to the west of the Pass, survives today, passing through the steep-sided glen cut by the River Garry.

Source: Clement Lemprière, *A Description of the Highlands of Scotland [showing] ye Forts lately Erected and Roads of Communication or Military Ways carried on by his Majesty's Command* (1731).

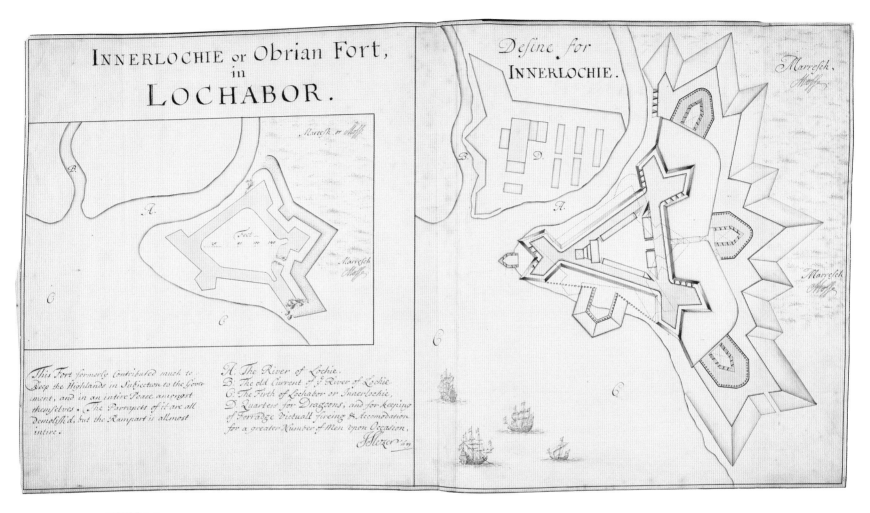

FIGURE 3.2

As a show of strength following the first uprising, Fort William (named in honour of the new king) was hastily built at the head of Loch Linnhe in Lochaber, a centre of Jacobite disaffection. The principal engineer in Scotland, John Slezer, drew up two plans for the fort. The first design retains the shape of Inverlochy Citadel, lying in ruins since the 1660 Restoration. The second shows a new, more extensive fort in place of the citadel, with three four-pointed bastions, ravelins and an extensive glacis (man-made slope), but this design was not adopted. Instead, the fort retained the shape of the original Cromwellian citadel: an irregular pentagon with a three-pointed bastion at the south-east and four two-pointed demi-bastions at the other corners. As Slezer notes 'This Fort formerly contributed much to keep the Highlanders in Subjection to the Government, and in an intire Peace amongst themselves', so presumably the government felt more elaborate fortifications were unnecessary. Built in only eleven days, the fort was little more than an earthwork with a palisade on top. Work continued on Fort William for the next sixty years.

Source: John Slezer, *Innerlochie or Obrian Fort, in Lochabor*; *Desine for Innerlochie* (1689). Courtesy of the British Library Board.

these actions came a change in mapping technologies, as engineers were commissioned to reconnoitre the Highlands, choose sites for new military establishments, and ensure effective communication between them and the principal Lowland garrisons (see chapter 4).

This chapter, and the next two, pay particular attention to the changing focus and nature of the military mapping of Scotland in the eighteenth century, and the imperatives behind the making of the maps. From the late seventeenth century, virtually all military engineering done in Scotland was undertaken through the Board of Ordnance, and many of the maps were produced by Ordnance personnel. The Board also acquired and made use of printed maps and maps produced by regular soldiers and overseers of the forts. Commemorative maps, some still in private collections, also contribute to the beautifully illustrated archive of military maps of Scotland in the eighteenth century. By examining the work of the Board of Ordnance in Scotland, and in considering the context of the maps' overall creation, distinction can be made between three main types. Firstly, maps made as part of the 'everyday' work of the military engineers – mostly fortification plans or road maps – and, secondly, the surveys and maps undertaken as specially commissioned projects, such as the Military Survey of Scotland (1747–55) and presentation maps. We may also note in this category battle plans. A third category of maps were those prepared as copies in training draughtsmen.

The Board of Ordnance

The origins of the Board of Ordnance can be traced to the fourteenth century when the Privy Wardrobe began to act as an itinerant armoury for the royal forces campaigning in Wales. By 1485, an Office of Ordnance was established; distinct and separate from the Wardrobe, it grew substantially during the reign of Henry VIII. In 1597, the Board was endowed with the responsibility for the upkeep and repair of forts and castles, in addition to supplying the army and navy with weapons, ammunition, equipment and stores.

In 1683, the Board was restructured and the functions of its civil and military officers were more clearly defined in a set of *Rules Orders and Instructions*. The engineers were answerable to the Surveyor-General, whose duty was to 'survey all Stores and Provisions of War' and the state of fortifications. He was charged with examining the qualifications and abilities of prospective engineers, all of whom were expected to be 'well skilled in all Parts of the Mathematicks, more particularly Stereometry, Altemetry and Geodasia, . . . in all manner of Foundations, in the Scantlings of all Timber and Stone, and of their several Natures'. The engineers were to take 'Surveys of Land, . . . to draw and design the Situation of any Place in their due Prospects, Uprights and Perspective'. Their knowledge of both civil and military architecture was to give rise to 'perfect Draughts of . . . the Fortifications, Forts and Fortresses of Our Kingdoms, . . . and to know the Importance of every one of them, where their Strength or Weakness lyes' (fig. 3.3). They were also to represent to the Board the necessary materials to be used along with their costs, instruct the master workmen in their jobs, supervise the building and design of fortifications, and conduct sieging operations.

In response to the changing nature of warfare and an increasing number of military operations and overseas garrisons, the number of engineers in the military establishment of the Board of Ordnance gradually increased. Ten engineers in 1699 increased to twenty-nine in 1745 (at least six in Scotland), thirty-seven in 1756 and sixty-one in 1759. The state's increasing dependence on specialist personnel also necessitated a change in military education – one based on formal training rather than simply experience. In 1741, the Royal Military Academy at Woolwich was established to instruct new recruits. The Academy did much to promote map-making as part of military culture. The curriculum included four groups of subjects: fortification and artillery; mathematics and geography; drawing to enable the topographic 'in-filling' of mathematical frameworks; and classics, writing and common arithmetic.

With a greater emphasis on mapping and maps for state purposes, there was also a need for a secure repository, one

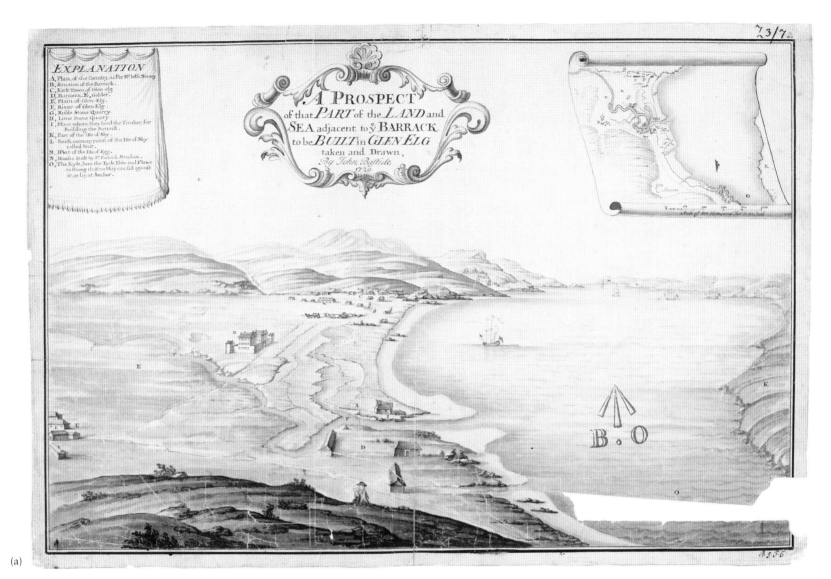

(a)

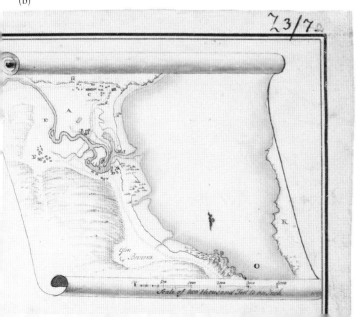

(b)

FIGURE 3.3

Bernera, near Glenelg in Wester Ross, was one of four barrack forts built in the 1720s for the purpose of accommodating soldiers to police the more remote parts of the Highlands.

(a) In the foreground of Bastide's scenographic view, an engineer can be seen at work, drawing what lay before him to the south-west, down the Sound of Sleat with the island of Skye to the right. In the top right corner is an unrolled topographic map of the environs (shown as an enlarged detail in (b)). In his view and map, the engineer has shown the site of the new fort and the layout of the buildings, as well as the fort's strategic situation, guarding the crossing to Skye. This was 'the art of depicting with a soldier's eye' the nature of the ground for military operations. The engineer was also required to provide detailed descriptions of the materials and their costs for any new works proposed to be done. With this in mind, Bastide has identified local resources: rubble and limestone quarries (G and H), and a place to land timber (I).

Source: John Henri Bastide, *A Prospect of that Part of the Land and Sea adjacent to ye Barrack to be Built in Glen Elg* (1720).

that afforded the Board and military personnel access to maps as well as a means to produce or duplicate them. Maps and plans were sent to a 'Drawing Room', part of the civilian establishment of the Ordnance based at the Tower of London, where they were classified according to geographical area and listed in a large folio volume, the 'Register of Draughts'.

The Drawing Room was, from the outset, a centre of carto-reproduction. Apprentices from the age of twelve were trained as draughtsmen and surveyors to complement the engineers in the field and to help with the compilation, drawing, correction, reduction, enlargement and copying of maps and plans. Copying was the most important part of a draughtsman's training, as well as an official requirement of the Ordnance, and was fundamental in establishing a consistent cartographic style founded on early Drawing Room practices. To this end, a style of British military mapping, influenced by continental practices, evolved. Types of representations – horizontal sections or plans, vertical sections or profiles, and bird's-eye views (fig. 3.4) – were of fundamental importance in structural and strategic processes. Colours used on plans and profiles

FIGURE 3.4

This striking bird's-eye view of Edinburgh Castle by John Slezer forms part of George III's topographical collections. During the sixty years of his reign (1760–1820), George III built up a considerable library of paintings, drawings, scientific instruments and printed books, as well as an extensive geographical collection. This comprised manuscript and printed maps and views, a few estate maps, military plans, maritime charts and 'topographical ephemera'. Slezer's drawing from the south shows the buildings within the castle as they were at the end of the seventeenth century, including the governor's residence, the storehouse, barracks, laboratory and guardhouse, and the parade ground (far left). Slezer has also drawn an elaborately fortified entrance to the castle, including an entrenched hornwork (a pair of projecting demi-bastions) below the round battery and counterguard (the defensive rampart in front of the castle), supported by multiple bastions. The design, which was never built, is based on a style of artillery works used in Europe by Sébastien le Prestre (1633–1707), Marquis de Vauban, the foremost military engineer of the seventeenth century and Louis XIV's general adviser on fortifications.

Source: John Slezer, A coloured bird's-eye view of Edinburgh Castle, showing the projected outworks (*c*.1690). Courtesy of the British Library Board.

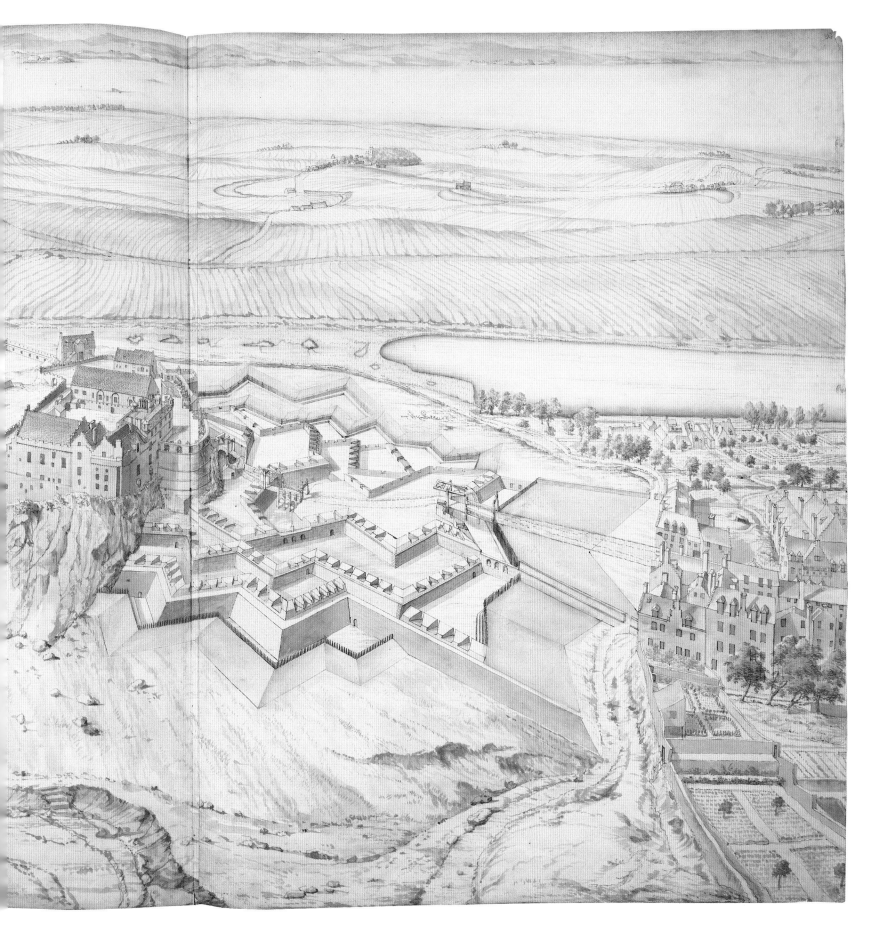

were also standardised. Indian ink was used to draw the outlines of most works. Masonry constructions were represented in red, elevations in a paler red wash, projected or incomplete works were washed in yellow, and turf-covered parapets and man-made inclines such as the glacis were green. The extent to which maps and plans consistently conformed to standard practices varied, however, often as a consequence of the conditions under which the original surveys were performed.

As with the establishment of engineers, the Drawing Room witnessed a growth in skilled map-makers as more maps were demanded by the state. Between 1720 and 1740, the staff increased from three to eight. In the period of the Seven Years' War (1756–63), the number increased to twenty-eight and by the second year of the American Revolutionary War (1775–83), the number had risen to thirty-six.

Fortifications 'for preventing an insult'

The Union in 1707 between England and Scotland led to widespread discontent within the Scottish population, which helped to reignite the Jacobite cause. With the support of Louis XIV of France, James VIII (and III of England), the Jacobite claimant to the throne, set sail from Dunkirk in March 1708, bound for Scotland. Despite being battered by storms, the thirty-strong fleet carrying 6,000 French troops cast anchor at the mouth of the Firth of Forth. Before James and his French allies could be set ashore to join with Jacobite lairds and their armed followers in Stirlingshire, the Royal Navy under Admiral Sir George Byng came in sight and the French ships fled north, eventually returning to Dunkirk.

The government had been lucky, for the defences of Scotland were in a very poor state. When the Privy Council requested information about Edinburgh and Stirling castles, the Board of Ordnance confessed that they had no maps of the castles, or of Fort William, and no estimates of what it would cost to put them 'in a posture of Defence'. The Board turned to their principal engineers in Scotland, John Slezer and Theodore Dury, who they believed could apprise them 'in less time, and at less charge' of the situation. Thereafter, maps and plans produced by Slezer and Dury were sent to London for consideration by the Board (figs 3.5 and 3.6).

The medieval fortifications inherited by the Board of Ordnance comprised castles and towers which had ceased to be defensively effective because of their ancient structures – most notably the castles of Inverness, Edinburgh, Stirling, Dumbarton and Blackness, and the military outposts of Castle Duart, Castle Tioram, Eilean Donan and Glengarry. The hope was that cheap and basic remedial works could update the castles to withstand methods of 'modern' warfare. Initial defensive works to Stirling and Edinburgh castles were proposed by Dury, who became Chief Engineer in Scotland from 1702. His plans were ambitious, designed 'for fortifying . . . to resist an Attack in form with Great Artillery', and proved too expensive for the Board. He resubmitted plans for works that would instead fortify 'for preventing an insult' but these still met with criticism (fig. 3.7). Work continued on Edinburgh and Stirling castles' defences through the first half of the eighteenth century and eventually proved to be highly effective. Neither castle fell during the 1745 Rebellion despite intense sieges from Jacobite armies.

The 1715 Rebellion and the military response

In 1714, Queen Anne died and was succeeded by George I of Hanover. The succession was controversial, and many Scottish nobles, already disaffected with the Union, were willing to rise in favour of a Stuart monarchy. In a letter to the Board of Ordnance in August 1715, Robert Johnson, overseer at Fort William, reported that the garrison was 'in great danger of being surprized by the Highland Clanns, they being all ready to rise, and they expect the Pretender to land every day'. Johnson was right to be concerned, as Fort William was precariously situated, both in terms of the immediate terrain (fig. 3.8) and its broader geographic location, surrounded by hostile Jacobite clans.

When John Erskine, Earl of Mar, who had served both William and Anne, was denied royal favour by George, he returned to Scotland and raised the Stuart Royal Standard at a gathering of the clans at Braemar on 6 September 1715. As the clans gathered, they successfully manoeuvred against the government garrisons, capturing soldiers at Glengarry, Castle Tioram and Eilean Donan as they marched south. The Jacobites quickly, and with little struggle, also took Perth (fig. 3.9), where a substantial Jacobite force, comprising 6,290 foot and 807 horse, formed under the command of Mar. Meanwhile, John Campbell, Duke of Argyll, gathered a substantially smaller but experienced government army of 960 Dragoons and 2,200 Infantry to stop Mar.

The armies met at the battle of Sheriffmuir, fought on Sunday, 13 November 1715. There are many eyewitness and participant accounts of the battle. However, one primary source is a contemporary Board of Ordnance plan detailing the engagement, of which there are three printed copies (fig. 3.10). These plans are unsigned but the Office of Ordnance 'Register of Draughts' attributes an original 'Plan of the Battle of Sheriff Muir' to Col. Lascelles. By the eighteenth century, battle maps were a well-established form of military narrative. Each campaign had its own distinctive character, as did the maps of the battles fought. The majority of the extant battle maps of the Jacobite campaigns in Scotland can be attributed to Hanoverian draughtsmen: military engineers such as John Henri Bastide, Dugal Campbell, John Elphinstone, William Eyres and Daniel Paterson, or draughtsman Thomas Sandby. A battle map's final form depended on the role it played in a military engagement, varying from simple pictorial plans of the order of battle, lacking both topographical details and scale (see chapter 4), to perspective maps and views showing the disposition of the opposing armies in relation to the surrounding topography, the whole drawn to scale.

Sheriffmuir had shown just how serious a threat the Jacobites were to the stability of the British state, and the surviving printed map was intended to commemorate the Hanoverian victory. While some saw Jacobitism as a political problem that should be resolved through political negotiations, the London government increasingly saw it as a military problem that required a military solution. Like Cromwell before them, they opted to establish a permanent military presence in the Highlands, building four new barrack forts (fig. 3.11) at Bernera near Glenelg (see fig. 3.3), Kiliwhimen in the Great Glen, Ruthven in Badenoch, and Inversnaid near the shores of Loch Lomond (fig. 3.12). One of the earliest military road maps was also drawn at this time in order to plan communications between the four barracks (fig. 3.13).

The construction of the barracks took years – Inversnaid, Kiliwhimen and Ruthven were ready to receive their garrisons in April 1721, Bernera in April 1723. Although intended to house between 120 and 360 soldiers, the barracks were never fully garrisoned. During his reconnaissance of the Highlands in 1724, Lieutenant-General George Wade reported that there were 'in some but thirty men'. He was particularly scathing about the effectiveness of the barracks: 'It is to be wish'd that during the Reign of Your Majesty and Your Successors, no Insurrection may ever happen to Experience whether the Barracks will Effectualy answer the End Propos'd'. His damning comments were directed more at the lack of troops than the design and strength of the fortifications: 'if the Number of Troops they are built to Contain, were constantly Quarter'd in them ... and proper Provisions laid in for their Support during the Winter Season, They might be of some Use to prevent Insurrections of the Highlanders', although they would never be strong enough to withstand a siege with heavy artillery.

The battle of Glenshiel, 10 June 1719

Not for the first time, the Jacobites soon found allies in Britain's enemies. While responding to Jacobite risings at home, Britain had also been engaged in the War of the Spanish Succession between 1701 and 1713. The war formally came to an end in 1713 with the Treaty of Utrecht, which saw Philip V succeed to the throne of Spain, but when Spain invaded Sardinia and Sicily in 1718, the British declared war against Spain. In retaliation, Philip V of Spain offered assistance to the Jacobites to

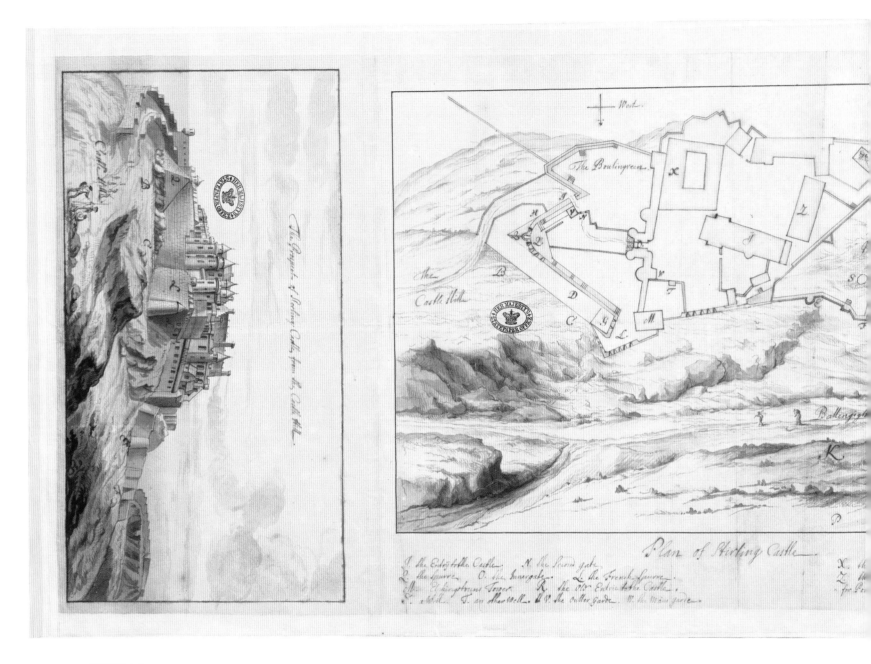

FIGURE 3.5

The location of Stirling Castle had been strategically important from the earliest times. Situated on a high volcanic rock, not only did the castle command one of the few bridges crossing the River Forth, but the river's floodplain below the castle afforded a convenient place for a military encampment of up to 12,000 men should the need arise. John Slezer notes that his plan, which is orientated with west at the top, and prospects, were drawn to accompany an account of the 'Defects of the Place', particularly its vulnerability to artillery fire from the direction of the town or Castle Hill. Slezer's prospect from Castle Hill also usefully shows us the French spur constructed in the 1540s, similar to that constructed at Edinburgh

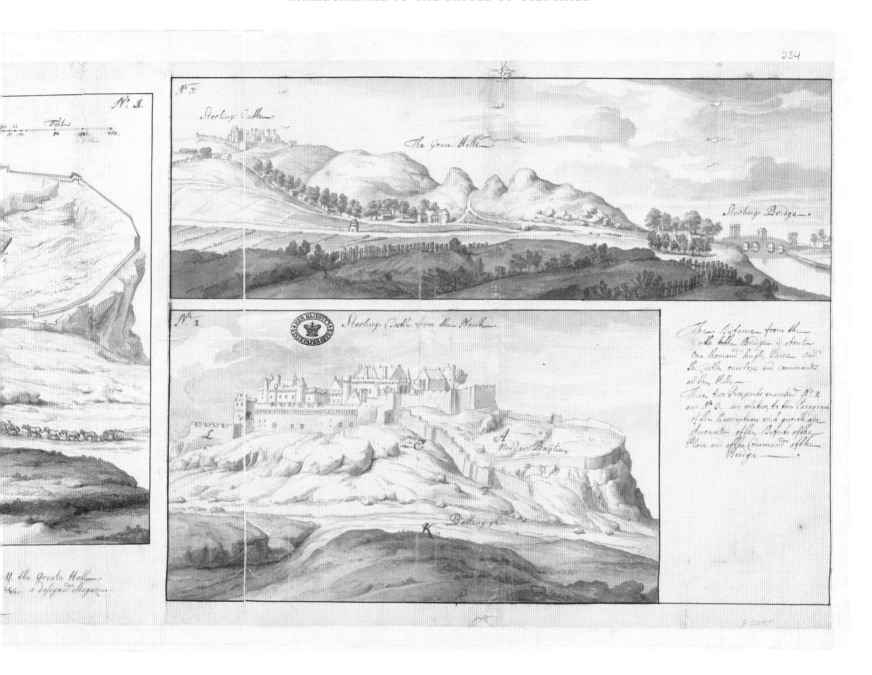

Castle (fig. 2.4). His focus on the view from the north and the medieval gatehouse, however, match later surveys that identified the walls on the east and north sides of the castle as old and too weak to withstand artillery fire. Between 1699 and 1703, improvements were made to the outer defences, which were then reconstructed during the period 1708–14 (see fig. 3.6).

Source: John Slezer, (1) *Plan of Stirling Castle*; (2) *Prospect of Sterling Castle from the North*; (3) *Prospect of Sterling Castle, the Goun [Gowan] Hills, Sterling bridge*; (4) *The Prospect of Sterling Castle from the Castle Hill* (c.1696). Courtesy of The National Archives.

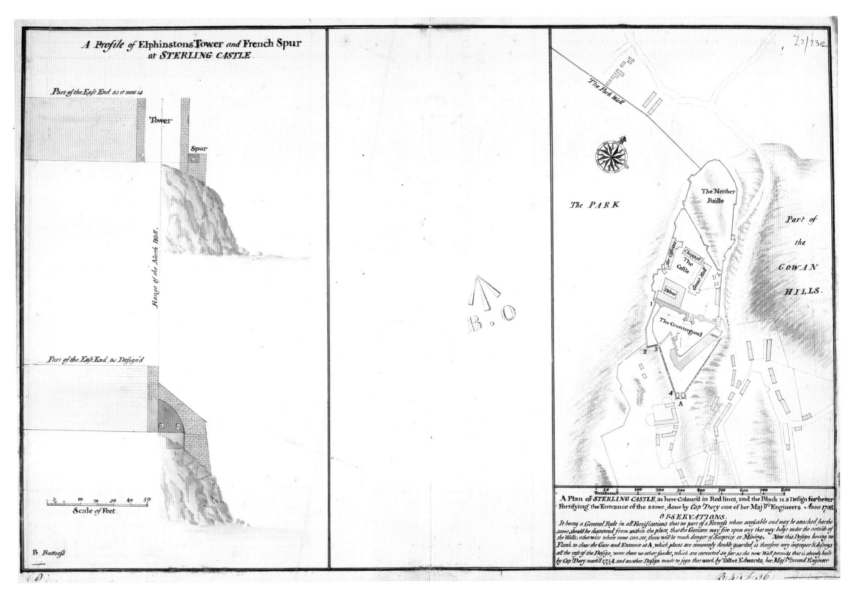

FIGURE 3.6

In 1708, Theodore Dury surveyed Stirling Castle and submitted plans for improving the entrance and towers near the counterguard (the defensive rampart facing the town), which were potentially exposed to gunfire from outside the castle, as well as providing shelter to any hostile force reaching the castle walls. Between 1708 and 1714, a new wall (marked 1, 2, 3, 4 on the plan on the right) was built to flank the gatehouse and entrance to the castle (marked A), the two central towers of the gatehouse were lowered and the side towers were reduced to their foundations, as shown by Dury's profile of Elphinstone's Tower (lying to the east of the counterguard). Dury's large-scale profile shows how the design of the tower was adapted to fit the situation of the castle – a skill required by all military engineers.

Source: Theodore Dury and Talbot Edwards, *A Plan of Sterling Castle, as here Colour'd in Red lines, and the Black is a Design for better Fortifying the Entrance of the Same; A Profile of Elphinstons Tower and French Spur at Sterling Castle* (1708).

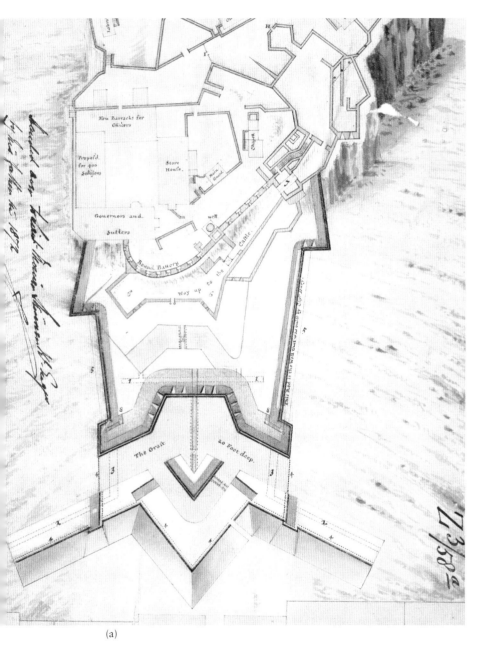

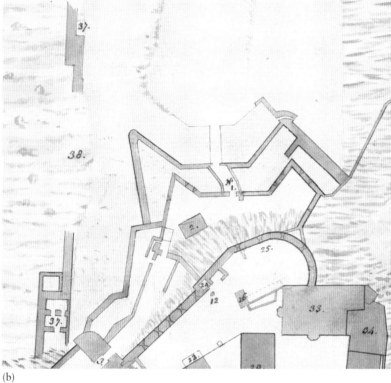

FIGURE 3.7

(a) This plan of Edinburgh Castle, orientated with west at the top, shows the proposed new artillery defences over what is now the Esplanade. For fortifying the entrance to the castle, Dury proposed a hornwork (a pair of projecting demi-bastions). In 1708, work began on the construction of a wall on the north side of the castle. The wall, shown here in red, was 200 feet long and 30 feet high. Dury's design, which he called 'Le Grand Secret', was criticised and revised in 1710 by Talbot Edwards, Second Engineer to the Ordnance, who found the hornwork too broad for Castle Hill. In its place, Edwards proposed a thinner and substantially longer hornwork, which could be accessed from the castle via small tunnels and defensive ditches. Edwards included a gallery from which to blow up the hornwork should it fall to an attacking force.

(b) Although Edwards' intervention effectively scuppered work on Dury's scheme, a lack of funds from the Board of Ordnance meant neither proposal was constructed. The more limited eastern defences finally built between 1714 and 1719 can be seen on this plan by John Romer in 1737. This also shows 'Le Grand Secret Designd by Monsr Dury' (37); a small section of Dury's wall survives today in Princes Street Gardens.

Sources: (a) Talbot Edwards, *A Plan of Edinburgh Castle* (1710).
(b) John Lambertus Romer, *A Plan of Edinburgh Castle* (1737).

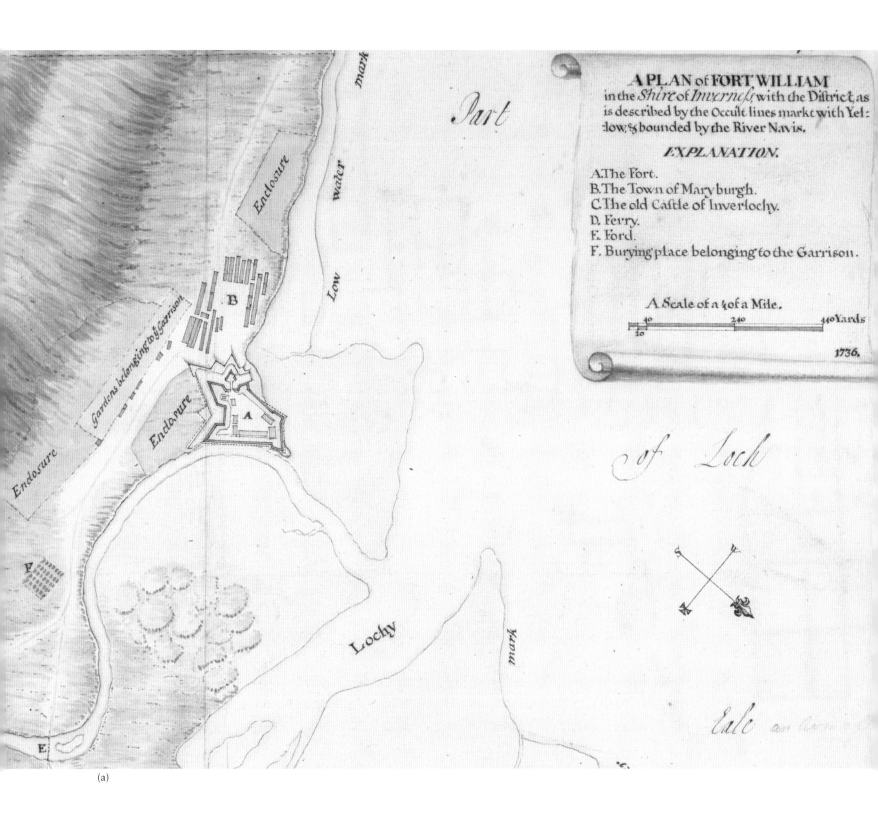
(a)

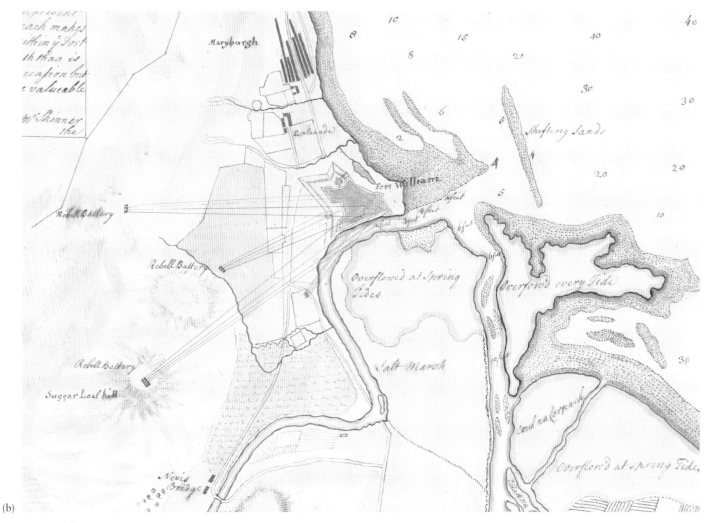

(b)

FIGURE 3.8

(a) (*Opposite*) This plan of the environs of Fort William emphasises just how vulnerable the fort was. While surveying the fort in 1710, engineer John O'Bryan (or O'Brien) found that every part of the fort was within musket shot of the surrounding mountains. Reviewing its situation later that same year, Talbot Edwards reported that 'This Forte for defending it self is indeed but a very ill figure', but its strategic location at the south-western end of the Great Glen meant that work continued on Fort William for the next sixty years. Once the earthen ramparts and bastions were faced with stone, attention turned to the construction of barracks and storehouses to accommodate a permanent garrison. The river and Loch Linnhe provided natural defences against ground attack from the north and west, and a covered way (a defensive passageway) and glacis (a bank sloping down from the fort, exposing the attackers to defensive fire) were built to protect the south and east flanks.

(b) (*Above*) In March 1746, when Fort William was under attack from Jacobite forces, the full extent of its disadvantageous position was revealed. On the landward side, the Hanoverian garrison was overlooked by four rebel batteries positioned on Cow Hill and Sugar Loaf Hill to the south-east and east of the fort. This extract from John Elphinstone's plan shows the situation of the batteries and their firing range across all areas of the fort. Despite its poor state, unfavourable position at the foot of a mountain and being under fire from the rebel batteries between 20 and 27 March 1746, Fort William successfully withstood the Jacobite siege – the only garrison fort in Scotland to do so during the 1745 Rising.

Sources: (a) [George Wade], *A Plan of Fort William in the Shire of Inverness* (1736). (b) John Elphinstone, *A Plan of the Grounds Adjacent to Fort William* (1748).

FIGURE 3.9

(a) (*Right*) Lewis Petit's plan shows the recapture of Perth after the 1715 Rising, during which time 'it was fortitifed and possess'd by the Rebells in Scotland'. As well as showing topographical features and the shape and extent of the town, the plan shows features of military interest, including the revetments (protective outer walls) and V-shaped redans (projections) built by the Jacobites as defences around Perth. The revetments extend to the old Cromwellian fort, now partly ruined, built in 1652 on the South Inch, with a small canal dug by the Jacobites to fill the moat with water.

(b) (*Below right*) This citadel overshadowing the town of Perth was never built, but Petit's plan illustrates the influence of European military theory in fortification and cartography into Scotland. His design is composed of a pentagon, containing barracks aligned parallel to the outer walls and forming the perimeter of an internal parade ground, with five bastions. Two of the citadel bastions overlook Perth while the remaining three look out over the surrounding countryside. Extensive outworks, including triangular ravelins and revetments, were designed to protect the fort from artillery fire, while its position on higher ground to the northwest made the fort far less vulnerable to newer, more mobile cannon fire. In its depiction, the new citadel was devoted entirely to a military presence, and in its form and situation it served the double function of defending itself from outside attack as well as offering a means by which to subjugate the local population. We can only assume that a lack of funds and priorities given to the network of forts further north in Scotland prevented this scheme from being developed.

Sources: (a) Lewis Petit, *A Plan of Perth with the Retrenchment made about it by the Pretenders Engineers* (1716). (b) Lewis Petit, *Plan of Perth and Adjacent Places with a Projection of a Cittadel* (1716).

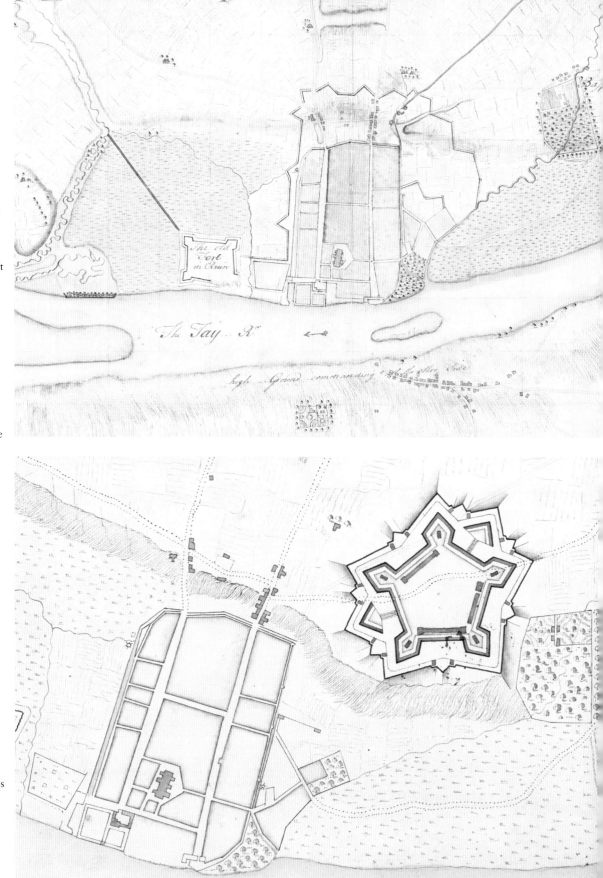

FIGURE 3.10

This map, orientated with east to the top, shows the positions taken by the Jacobites leading up to the battle of Sheriffmuir. On the night of 12 November 1715, Argyll's Hanoverian army encamped at 'Newtoun' near Dunblane. Mar's army gathered near Kinbuck and set up camp on either side of the road from Dunblane to Perth (position B). The following morning, Mar ordered his troops to form into four columns and directed them onto Kinbuck Moor (position C). What followed is explained on this northern section of the map. Unfortunately, the critical southern section (or right-hand side), which would have shown the main battlefield, is missing. By all accounts, the battle was a chaotic affair, but strategically, the Jacobites had lost the initiative to break past Stirling and into the Lowlands. From its pro-Hanoverian narrative, it seems likely that this map was specifically printed to commemorate the Hanoverian victory. The map shows the battlefield terrain, distinguishing tilled fields from open pastures, ridges of higher ground with the use of cross-hatching, and various sizes of settlements by way of three-dimensional houses.

Source: Anon., Plan of the battle of Sheriffmuir (1715).

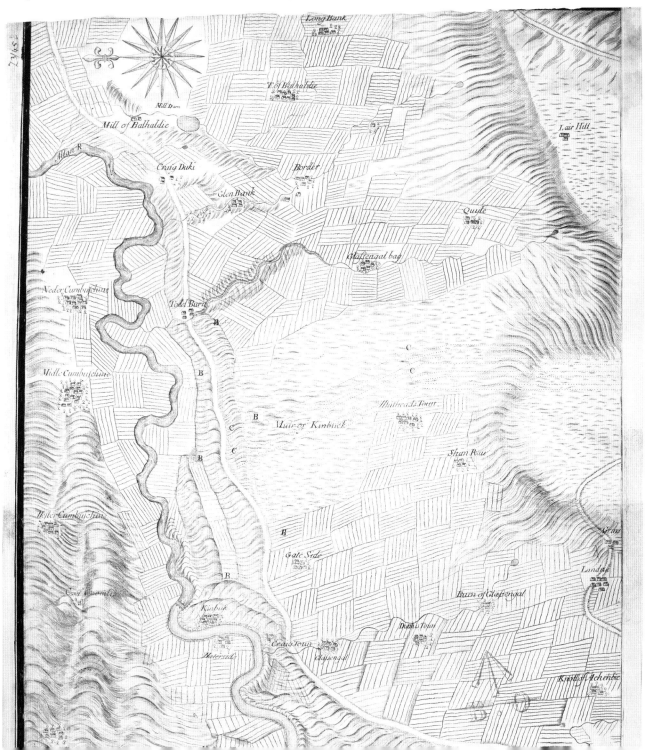

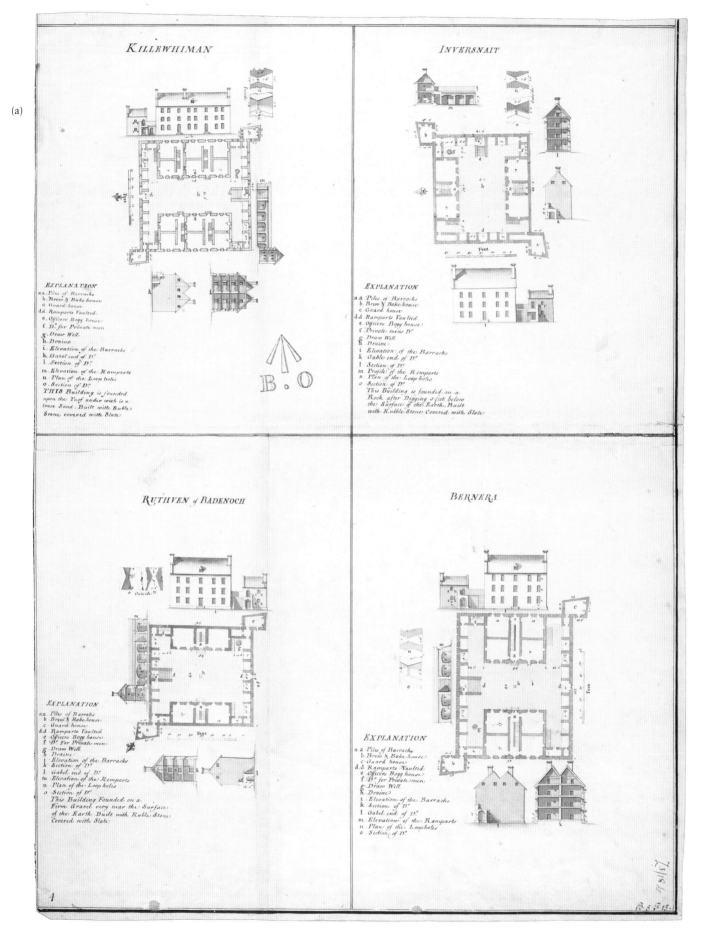

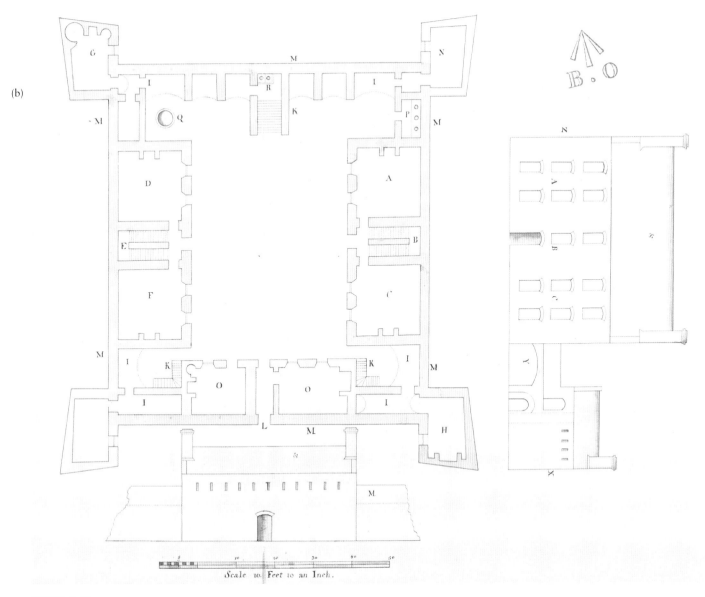

FIGURE 3.11

(a) (*Opposite*) These four sites – Kiliwhimen in the Great Glen, Inversnaid near the shores of Loch Lomond, Ruthven in Badenoch, and Bernera near Glenelg – were chosen by Lewis Petit in 1717 as suitable sites to build new 'barrack forts'. The design of these forts was of particular concern to military engineers, a fact reflected in the numerous large-scale site plans made by Andrews Jelfe, Architect to the Ordnance from 1719 to 1727. Each took a similar, but not identical, architectural form to the others. Each stood as a detached, enclosed, self-defensible barrack complex with bastion-like angle towers at alternate corners from which it was possible to cover the whole exterior of the enclosure with flanking musketry fire.

(b) (*Above*) The original plans had also included a tower at each corner to provide complete enfilading (sideways) fire to the entrance and sally-ports, but a lack of funds caused the number to be reduced to two. The towers were used as guard rooms, store rooms, bake- and brew-houses and the officers' quarters. Two sets of three-storey barrack blocks were built facing across a square with the rear walls forming part of the external defences.

The cost for building the four barracks was estimated at £9,300 in 1717. An advertisement was placed in the *London Gazette* and in newspapers printed in Edinburgh for artificers to tender for work on the barracks; interested parties were allowed to see the plans in order to propose costs which, once agreed by the Board of Ordnance, were binding. Fiscal restrictions meant that 'Those that would do it best & cheapest were entitled to work according to the Tenour of the Advertisement in the *London Gazette*.'

Sources: (a) Andrews Jelfe, *Killewhiman, Inversnait, Ruthven of Badenoch, Bernera* (1719). (b) Andrews Jelfe, *Innersnait* (1719).

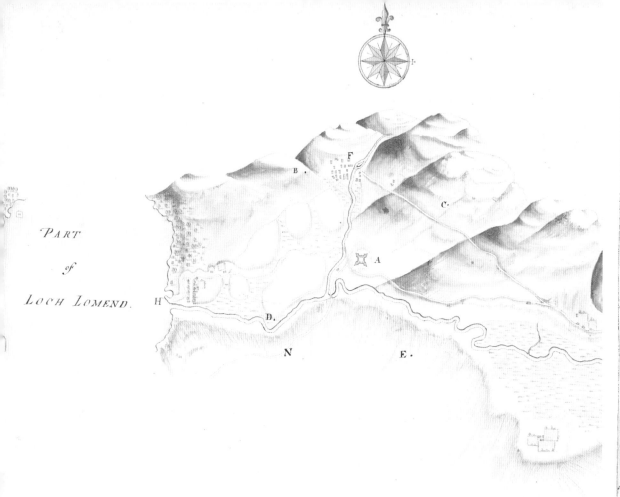

FIGURE 3.12

(*Above*) This map shows the area around Inversnaid barracks. The construction of the new barracks took several years, beginning with the protracted process of passing Acts of Parliament for 'investing the ground in the Crown' so that no civilian buildings could be built within musket shot of the intended barracks. This is marked on Dumaresq and Bastide's map as 'B.C.E.D. The four Angles of ye bounds about the Barrack being 704785 yards 5 feet square Including ye spot on [which the] Barrack stands'. The barrack (at position A) is shown with four towers (see fig. 3.11b). Until the barracks were completed, they were vulnerable to attack, and incidents of workmen being kidnapped were reported to the Board with an appeal for greater security. Rumours of rebel attacks effectively stopped work on Inversnaid until soldiers were put in place to protect the workmen (position F on the map indicates 'Innersnait huts possess'd by a Detachment of hundred men'). A letter from the commander of Inversnaid dated 21 April 1719 warned of Jacobites rising in support of the exiled Stuarts.

Source: John Dumaresq and John Henri Bastide, *A Draught of Innersnait, in the Highlands of North Brittain, nere the Head of Loch Lomend with Part of the Country Adjacent* (1719).

FIGURE 3.13

(*Right*) This map, orientated with west to the top, shows the routes between the new barracks in Scotland, and the very limited topographic knowledge of the wider terrain. As two of the overseers of the four new barrack forts being built in the Highlands from 1718, John Dumaresq and John Henri Bastide were concerned to plan communications between them. Together they surveyed the roads from Inversnaid on Loch Lomond (top left) to Ruthven Barracks by Kingussie in Strath Spey (bottom right) via Balquhidder, Glen Beich, Kenmore, Loch Tummel, Blair Castle and the Minigaig Pass, to Kiliwhimen at the southern end of Loch Ness (centre right) via the Corrieyairack Pass, and then south to Fort William (top right). The 'roads' at this time were little more than tracks, unsuitable for artillery or wheeled carriages. This deficiency, combined with the development of more mobile military strategies, gave rise to the planning and building of a new road system in Scotland following the arrival of George Wade in 1724 (see chapter 4).

Source: John Dumaresq and John Henri Bastide, *The Roads between Innersnait Ruthvan of Badenock Kiliwhiman and Fort William in the highlands of North Brittian* (1718).

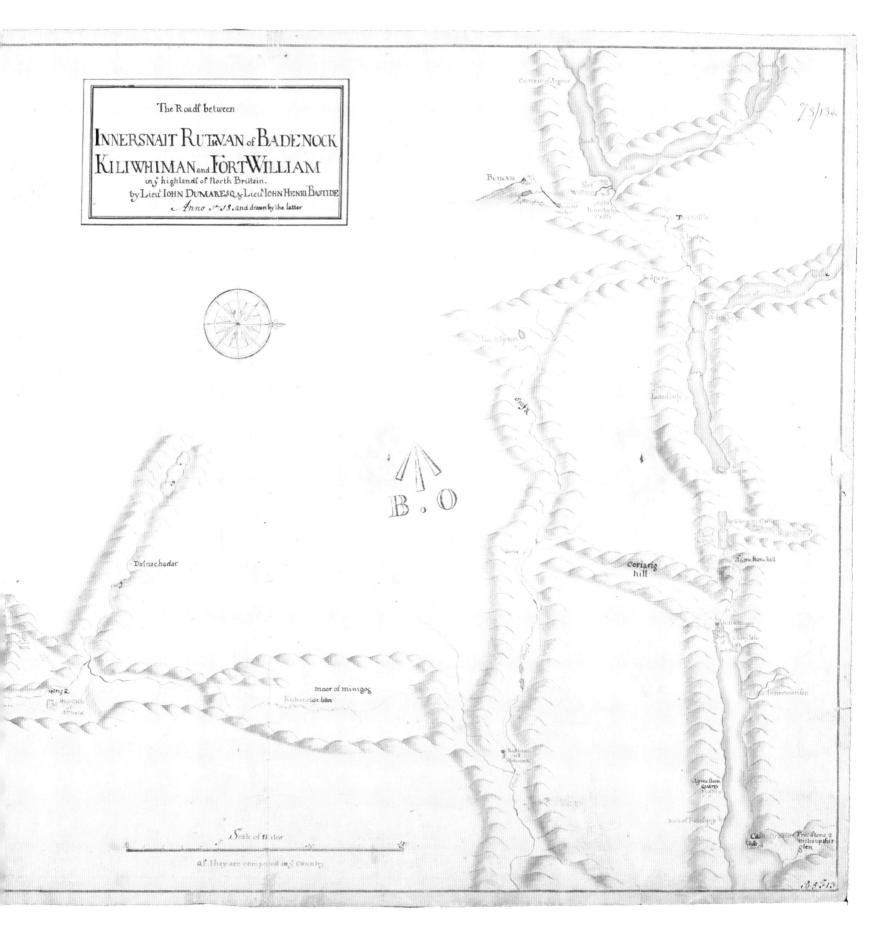

restore James Stuart to the throne of Britain. As well as plans to invade the south of England, the Spanish invasion plan included inciting rebellion in the Highlands.

When George Keith, the exiled Earl Marischal, landed at Stornoway/Steòrnabhagh on the island of Lewis in late February, he was accompanied by 300 Spanish infantry. There he was joined by Jacobites who had taken refuge in France, including Keith's younger brother, James, the Marquis of Seaforth, Lord George Murray and his older brother the Marquis of Tullibardine, under whose leadership they acrimoniously united. From there, they moved onto the mainland and established their headquarters at Eilean Donan Castle (fig. 3.14), situated on a small island at the head of Loch Alsh. The intention was to unite with the Highland clans under Robert Macgregor Campbell – better known as Rob Roy – and march upon Inverness where they were to take possession of the garrison, but from this point the plan started to unravel. With news of a naval force searching for insurgents along the west coast, Tullibardine led his forces away from Eilean Donan. At the same time, Major-General Joseph Wightman deployed his Hanoverian army from the garrison at Inverness. With a naval force threatening from the west and Wightman's army fast approaching from the north-east, the Jacobites had little choice but to make a stand and elected to face the Hanoverian army at Glenshiel in Wester Ross, a natural bottleneck where the Inverness road passed through mountains on either side.

Four almost identical contemporary manuscript maps of the battle of Glenshiel survive. All are drawn by John Henri Bastide in Indian ink and watercolour wash (fig. 3.15). Both sides were evenly matched, and fighting lasted several hours, but came to be seen as a victory for the Hanoverian army. The Spanish were left to surrender while the remaining Highlanders dispersed. Lord Carpenter, Commander in Chief of the Hanoverian forces, made no reprisals in response to the 1719 Rising other than to seize a great quantity of weapons from the rebels; defeat was enough for the Jacobites. Attention did turn, however, towards the west-coast defences. Andrews Jelfe, overseer of military works in Scotland, was ordered by the Board of Ordnance 'to loose no time in proceeding on the works at Bernera' in Wester Ross 'for keeping a Comunication with ye same'. The Board refused to continue work on the defences at Inverness on the grounds that they had run out of money (fig. 3.16). An engineer was instead to report on the recent works and what else was required to secure the town 'lying so near that part of the Highlands, where Rebells can with most safety assemble'. One further symbolically significant consequence of the 1719 Rising was the conversion of the royal chambers in Edinburgh and Stirling castles to barracks and gunpowder magazines (fig. 3.17).

In addition to the battle of Killiecrankie, with which we began this chapter, Clement Lemprière's military map of Scotland (fig. 3.18) shows the prominent Roman Road leading north through Annandale and the Beattock Summit. The map offers an important insight into the conquering Hanoverian mindset, blending past and present military infrastructure and events with past and present friends and enemies. How should these distant lands on the edge of the Empire be best conquered and absorbed? Over the past three decades, the military engineers in Scotland had taken the Jacobite threat seriously and done their best to plan major upgrades to existing castles and sometimes ambitious new forts, but continually found their proposals rejected or whittled down by the miserly Board of Ordnance in London. If the Hanoverians were not going to be forced to retreat to the Hadrianic frontier further south, they needed a new strategy and, even better, one backed up by real money. Our next chapter considers whether such a new strategy was actually implemented.

FIGURE 3.14

Eilean Donan Castle is situated at the head of Loch Alsh, a long sea loch opposite the Isle of Skye. In 1719, the castle was defended by forty-six Spanish soldiers who were supporting the Jacobites. The Jacobites had established a magazine of gunpowder at the castle and were waiting for weapons and cannon to be delivered from Spain. When news of the rising reached London, the government sent three Royal Navy ships to Loch Alsh to deal with the situation. After three days bombarding the castle, a storming party was sent ashore and succeeded in overwhelming the Spanish soldiers. The party captured the magazine and used the powder to blow up what remained of the castle. For 200 years, Eilean Donan Castle remained in ruins, but from Lewis Petit's plan and profile, we can see how the castle must have looked before it was destroyed.

Petit's plan is relatively plain, drafted in pen and Indian ink with a grey watercolour wash, quite unlike some of the highly coloured plans produced by other military engineers and draughtsmen. Simple executions such as Petit's emphasise the utilitarian nature of graphic representations. His use of a multi-perspective display provided essential military information at a glance, namely the establishment's form and layout, sometimes its situation, in order to provide a record of its defensive state.

Source: Lewis Petit, Plann of the Castle of Island Dounan; Profile of the Front of the Castle of Island Dounan... (1714).

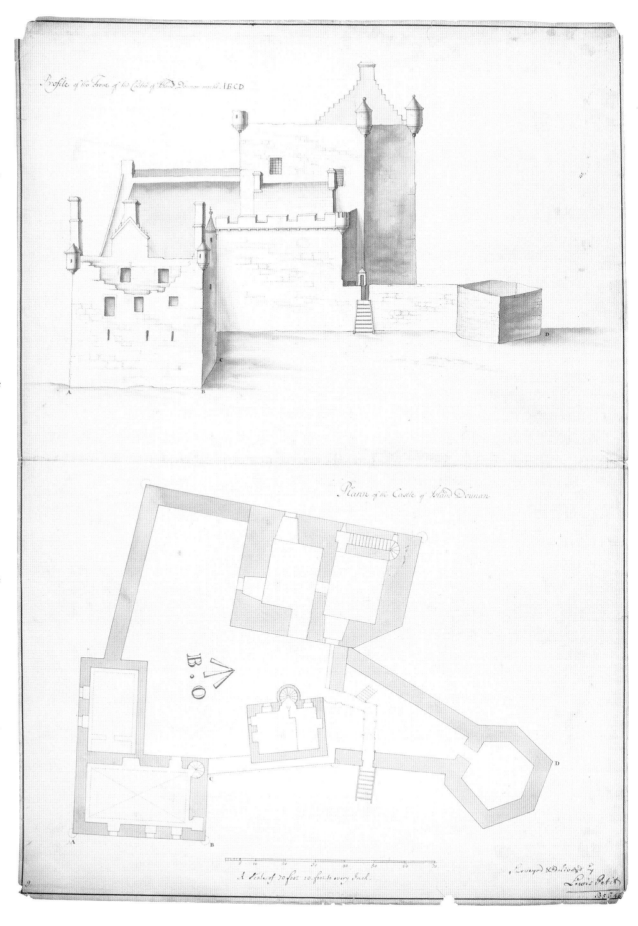

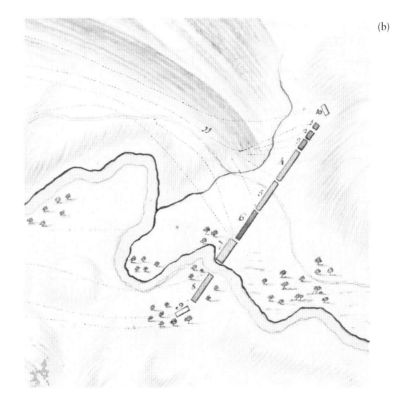

FIGURE 3.15

(a) The strategic value of the narrow mountain pass of Glenshiel is clearly shown in Bastide's scenographic map, which combines vertical and bird's-eye perspectives to portray the impressive relief of the glen. His plan of the field of battle offers an animated view of the action. Tullibardine's Jacobite forces, aided by a Spanish regiment, took up a defensive position on the imposing high ground to the north and south of the road to Inverness. Wightman, to the east, used mortar teams to bombard the Jacobite position before ordering his infantry to attack their flanks, while continuing to shell the Jacobite centre and Spanish troops. After three hours of resistance, the Jacobites were eventually driven from their defensive position and forced to retreat. The Spanish, meanwhile, made an orderly retreat and later surrendered.

(b) Bastide uses symbols, colour, shading, letters and numbers on the face of the plan to show the position of the armies and their various detachments during the course of the battle. Reflecting tactical theory, he distinguishes between infantry and cavalry. The rectangles for infantry are narrow and varied in length depending on the size of the unit: 'a Serj^t and twelve Granadiers' (unit 1) is depicted as a small square on the map, whereas 'Col. Montagu's Regim^t' (unit 4) is a long, thin rectangle reflecting the greater number of men. The Dragoons (unit 7) are shown with a broader rectangle. Colour was an essential aspect of the convention system. The cartographer had to ensure that the locations of the different armies were clearly distinguished. On Bastide's map, the British units of Wightman's army are shown in red and the Dragoons in green, the Dutch in blue, and Munro and Sutherland's Highlanders in white. The initial deployment of the Jacobite army is shown conventionally with rectangles coloured in yellow and grey. Lines of movement across the map are shown by dotted lines.

Source: John Henri Bastide, *A Plan of the Field of Battle . . . at the Pass of Glenshiels in Kintail North Britain . . .* (1719).

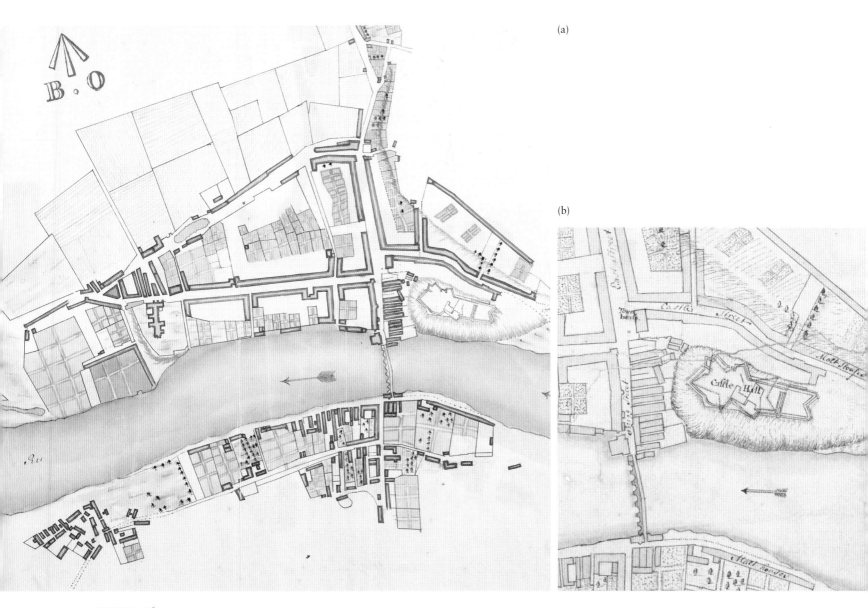

FIGURE 3.16

(a) This is a copy of one of the earliest surviving town plans of Inverness and clearly shows the seven-arch bridge, erected in 1685–89, across the River Ness. The original plan, from which (b) is an extract, was drawn by Lewis Petit during a reconnaissance of the Highland forts; the copy is in another hand and drawn at a later date, when military colour schemes had become more established. As for his plan of Perth (fig. 3.9a), Petit simply depicts the essential military topography of the town – its built-up areas, main river, open spaces, streets, communications and defensible sites – to help plan the future fort.

(b) In the original plan, Petit has drawn a red dotted outline of an irregular fort over the top of the medieval castle (Castle Hill). While some modifications were made to the old castle in 1718 (and in the 1720s and 1730s), the final form did not match Petit's proposal (see chapter 4).

Source: Lewis Petit, *Inverness in North Brittain* (1716).

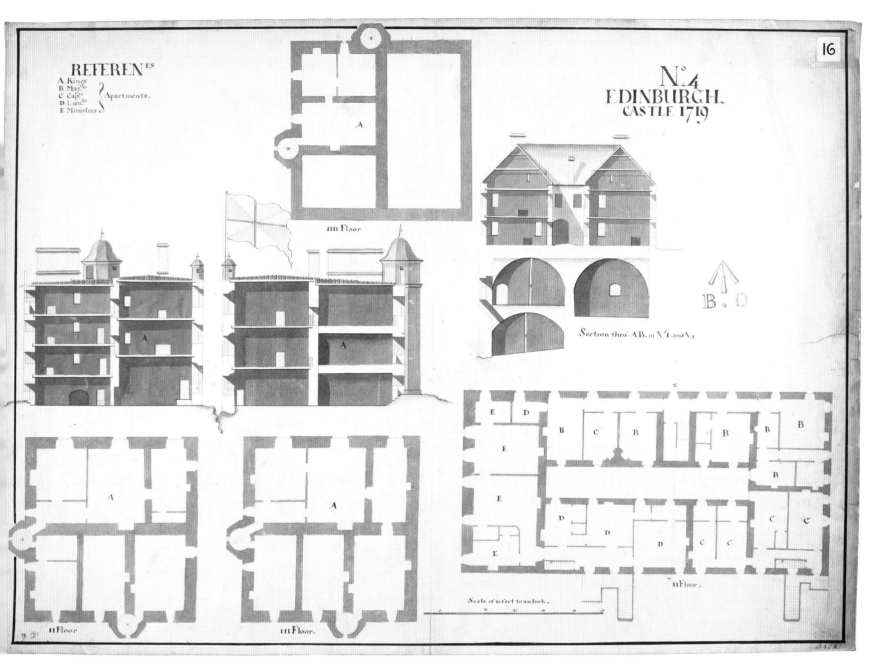

FIGURE 3.17

(a) This is one of a series of sheets showing floor plans and sections of different parts of Edinburgh Castle. They are mostly drawn in Indian ink, and show the redevelopment of the Royal Palace (marked A on this plan) and new accommodation for officers and gunners (marked B–E, and now known as the Queen Anne Building) in response to the failed Jacobite Rising of 1719. These two buildings stand facing each other on either side of Crown Square, at the heart of the castle.

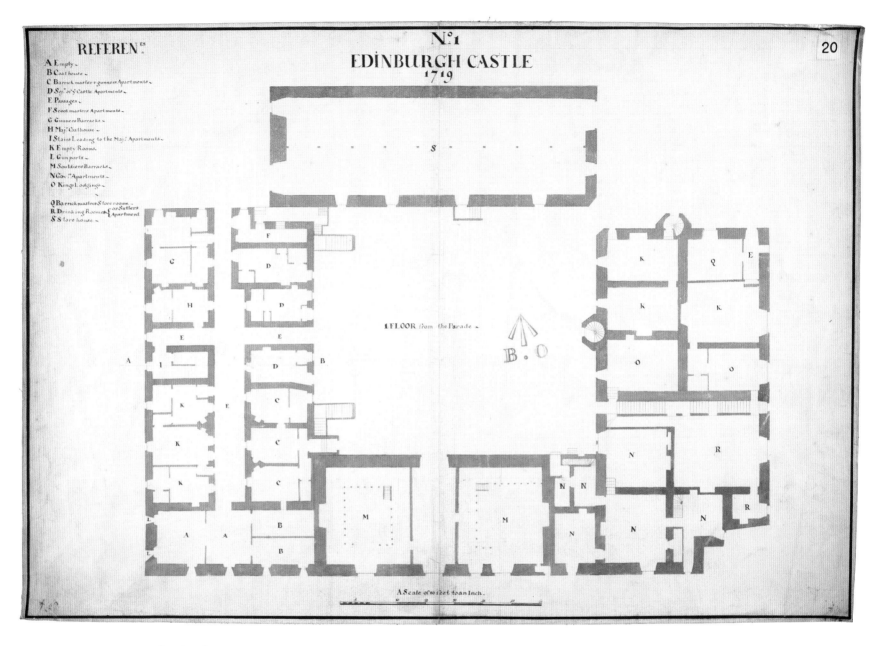

FIGURE 3.17 (*continued*)

(b) This plan shows how the construction of the Queen Anne Building to the west and storehouse to the north allowed the creation of a central courtyard. Although the primary purpose was new accommodation, the barracks also had an artillery function, and six new gunports were added (marked as L on this plan): four in the south wall, and two facing west over Dury's Battery. Also note how the profile of the Queen Anne Building (a) shows the underground vaults, which would see new uses almost a century later (fig. 6.5).

Sources: (a) Thomas Moore and Andrews Jelfe, *No. 4, Edinburgh Castle* (1719). (b) Thomas Moore and Andrews Jelfe, *No. 1, Edinburgh Castle* (1719).

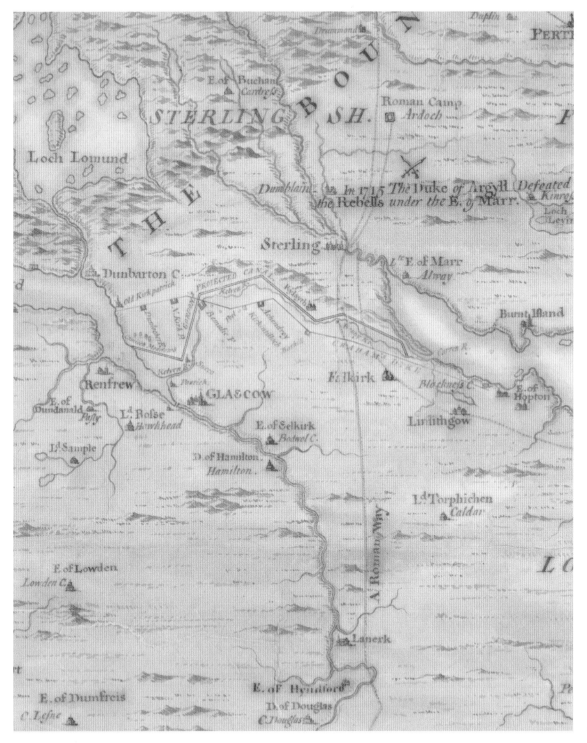

FIGURE 3.18

Clement Lemprière's military map of Scotland was drawn in 1731 to show the new military roads being constructed in the 1720s and 1730s, and the principal government garrisons in relation to the disposition, loyalties and strengths of the Highland clans. Lemprière also chose to show the prominent Roman Road, crossing the Antonine Wall ('Graham's Dyke') between the Forth and Clyde, and continuing north, via the Roman Camp at Ardoch. The Hanoverians evidently saw connections between the Roman conquests of Scotland and their own military strategies.

Source: Clement Lemprière, *A Description of the Highlands of Scotland [showing] ye Forts lately Erected and Roads of Communication or Military Ways carried on by his Majesty's Command* (1731).

FIGURE 3.19

(*Overleaf*) John Dumaresq and John Henri Bastide, *A Draught of Innersnait, in the Highlands of North Brittain, nere the Head of Loch Lomend with Part of the Country Adjacent* (1719).

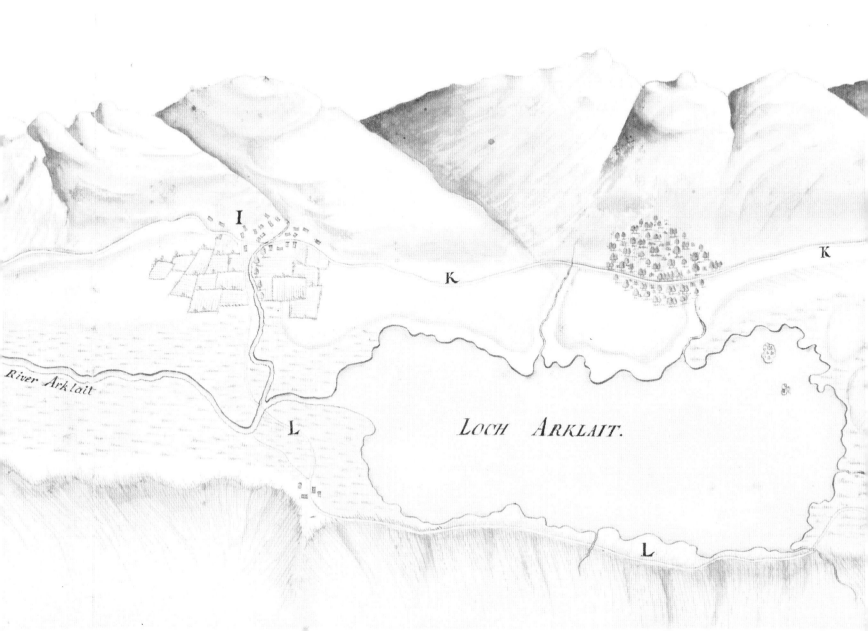

CHAPTER FOUR
GEORGE WADE TO THE BATTLE OF CULLODEN, 1724–46

In 1724, George I received a disturbing account from Simon Fraser, 11th Lord Lovat (fig. 4.1), 'concerning the state of the Highlands'. Lovat claimed that:

> The Highlanders . . . grow averse to all notions of peace and tranquillity, – they constantly practise their use of arms, – they increase their numbers, by drawing many into their gang who would otherwise be good subjects, – and they remain ready and proper materials for disturbing the government upon the first occasion.

The Disarming Act (1716) and the building of barracks and garrison forts, all designed to curtail Jacobitism in the Highlands, seemed to have failed. In July 1724, Lieutenant-General George Wade was despatched to Scotland with orders to ascertain how much truth there was in Lovat's memorandum

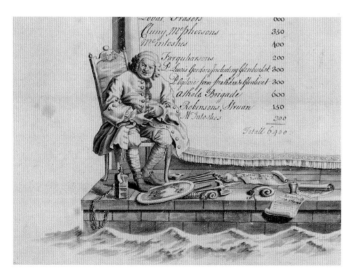

FIGURE 4.1

Guided solely by self-interest, the chief of Clan Fraser, Simon Fraser, Lord Lovat, switched his allegiance back and forth between the government and the Jacobites throughout his life. Eventually, in 1746, he was outdone by his own duplicity and poor judgement, having brought out 600 Fraser clansmen to fight for Charles Edward Stuart, the Young Pretender, at Culloden. Captured after Culloden, Lovat was executed at Tower Hill in 1747. Lovat is pictured here counting the numbers of the clans who supported Charles Edward on his fingers (based on William Hogarth's 1746 etching). The image forms part of an ornate cartouche listing the Highland Clans in the 1745 Rising (for a full list, see fig. 4.3b), which forms part of a map of Scotland drawn by engineer John Elphinstone. Elphinstone was commissioned to draw the map by William Keppel, Earl of Albemarle, to mark his appointment as Commander in Chief of the forces in Scotland in July 1746.

Source: John Elphinstone, *A New Map of North Britain* . . . (1746). Courtesy of the British Library Board.

and to make recommendations on how to address the problems. Wade's subsequent reconnaissance of the Highlands exposed a fundamental flaw in the nature of military planning in Scotland: it was too static. During Wade's sixteen years in Scotland, new measures were established to police the Highlands, to more effectively connect the Highland forts with each other and the Lowland castles, and to speed up the transport of munitions and stores. This generated a need for different maps – for surveys and descriptions of Scotland to assist military movement.

This chapter describes the changes in mapping coincident with the government's evolving approach to the problem of Jacobite unrest in Scotland. Between 1724 and 1746, military maps were drawn with three main functions in mind. The first was strategic planning – maps were drawn to plan routes and to record the planned spread of military access (fig. 4.2). The second was engineering, not just for new forts and castle defences, but also for building new roads and bridges. The nation's defence depended on the quick movement of troops and this required roads and bridges suitable for men and machinery to access remote parts of Scotland. The third was for tactical military manoeuvres in preparation for and during battle. Events during the 'Forty-five are portrayed in contemporary maps. Detailed battle plans of Prestonpans, Falkirk and Culloden provide dynamic but highly selective insights into these military encounters. Less detailed maps were also needed to plan and show the movement of armies across the country. The different audiences – military or civilian, Hanoverian or Jacobite – intended for these two types of mapping fundamentally influenced their content.

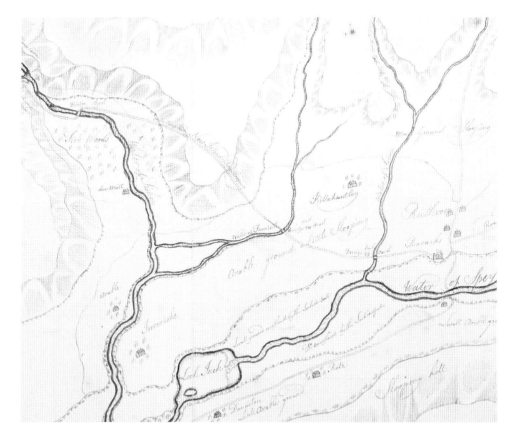

FIGURE 4.2

Proposals to build a new road from the barracks at Ruthven to Aberdeen, 'which would have open'd a Communication from the East Coast into the Highlands', were never fulfilled. Joseph Avery compiled this map of the 'New Intended Road' through the Cairngorm Mountains, from Ruthven to Braemar, via Glen Feshie and Glen Geldie. He used extracts from surveys he had made for the Duke of Gordon and also information from Major Caulfeild (William Caulfeild was in Scotland from at least 1729 and was appointed Inspector of Roads in 1732). Avery's map includes detailed remarks on the nature of the country and also notes the span of each bridge necessary to cross the numerous tributaries of the rivers Dee and Spey; this detail shows part of the River Feshie, which flows north to meet the Spey at Kincraig.

Source: Joseph Avery (and George Wade), *A Plan of the Country where the New Intended Road is to be made from the Barack at Ruthven in Badenoth to Invercall in Brae Marr . . .* (1735).

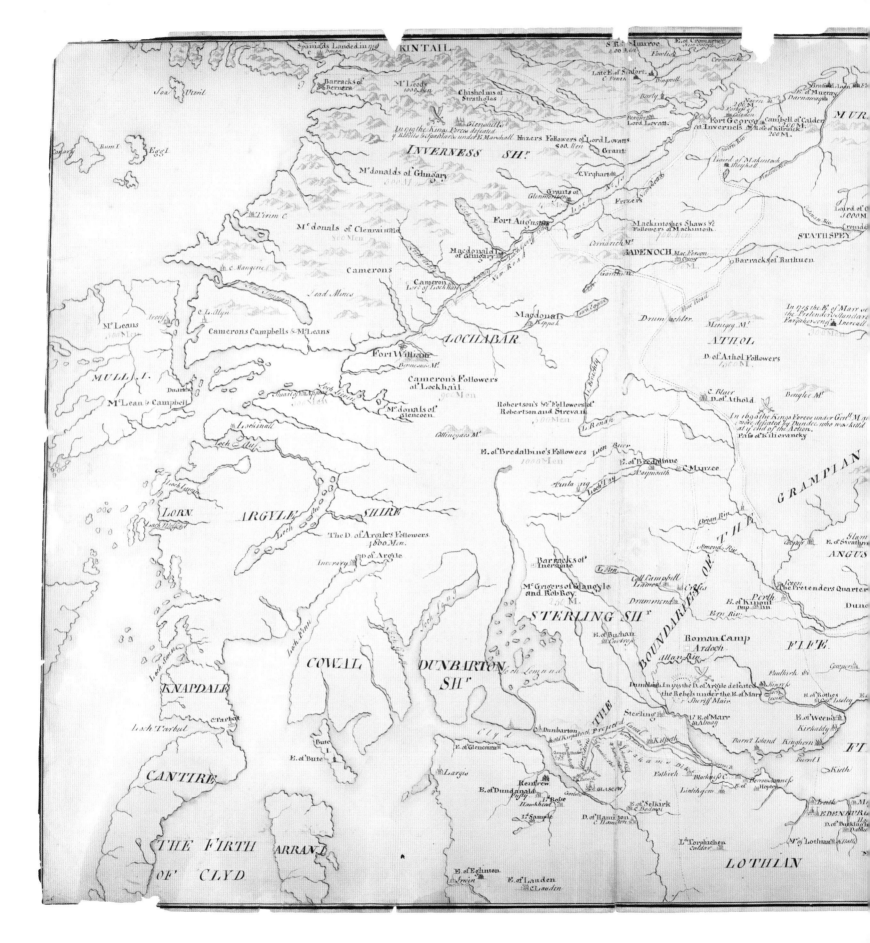

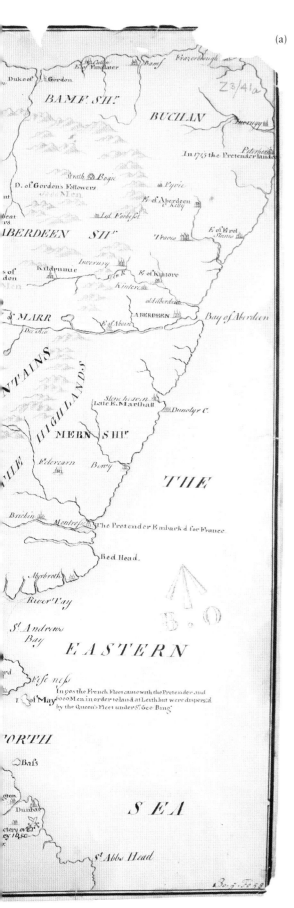

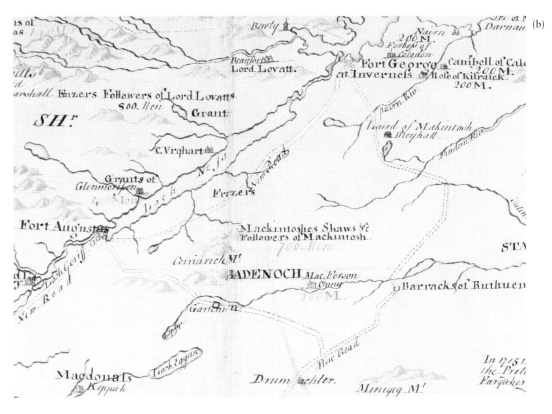

FIGURE 4.3

(a) This is one of several maps of the Highlands prepared for intelligence purposes, including information gathered by Wade during his reconnaissance of the Highlands in the 1720s. In line with Wade's proposals for bringing order to the Highlands, the map also shows the new and projected roads.

(b) As shown in this detail, the map indicates the clan territories by name, with the number of men each clan contributed to the 1715 Rising – Jacobites in red and Hanoverian supporters in black.

(c) As the *Explanation* accompanying the map shows, the Jacobites outnumbered the Hanoverian supporters, and the government's concern was that they would rise again.

Source: John Manson, Map of part of Scotland, showing clans that rebelled in 1715; *Explanation* (1731).

Strategy and engineering:
Wade in Scotland, 1724–40

On 3 July 1724, George Wade was instructed to 'go into the Highlands of Scotland, and narrowly to Inspect the present Situation of the Highlanders... and the state of the Country'. He was also to 'Suggest to Your Majesty, such other Remedies as may Conduce to the Quiet of Your Faithfull Subjects, and the good Settlement of that part of the Kingdom.' Wade wasted little time and, on 10 November 1724, he submitted his first report to the king. There was much to alarm the government in Wade's report, two entries in particular. The first related to the loyalty of the Highlanders. Of the 22,000 men in the Highlands capable of bearing arms, only 10,000 were found to be favourably disposed towards the government. Some contemporary maps record even fewer numbers supporting the government and considerably more in support of the Jacobites (fig. 4.3). These men had borne arms for the Jacobite cause in 1715 and many seemed ready to rise again, in favour of Charles Edward Stuart, the Young Pretender. The second entry was news that efforts to disarm the clans after the 'Fifteen had been:

> so ill Executed, that the Clans the most disaffected to Your Majestys Government, remain better Arm'd than Ever, and Consequently more in a capacity . . . to be used as Tools, or Instruments to any Foreign Power or Domestick Incendiarys who may attempt to disturb the Peace of Your Majestys Reign.

In May 1725, Wade was made Commander in Chief of all the king's forces, castles, forts and barracks in Scotland as a consequence of making several recommendations for establishing order in the Highlands. These included forming independent companies of locals to police the Highlands and enforce the Disarming Act. Wade also proposed building two new forts. The first was a new barrack fort at Inverness. The medieval castle on Castle Hill was no longer effective as a fortification (see fig. 3.16), but its commanding situation within 150 yards (137 metres) of the bridge across the River Ness made it strategically important (fig. 4.4); the bridge was the only crossing between the north and south Highlands for 30 miles (as far as Kiliwhimen in the Great Glen). Wade was not a military engineer, but he knew the value of accurate surveys and maps to plan new infrastructure for controlling a hostile population. To this end, Wade commissioned John Lambertus Romer to survey the castle in 1725 with the purpose of converting it into barracks (fig. 4.5). Romer had been appointed engineer in charge of the works and barrack building in Scotland in 1720, replacing Andrews Jelfe, and proved to be one of the Board of Ordnance's finest military engineers. He learned the art of engineering from his father – Wolfgang William Romer – a Dutch military engineer who served under William, Prince of Orange, and accompanied him to England in 1688. Wolfgang Romer was overseer of the works at Albany, New York, and Portsmouth, where he was assisted by his son. In 1710, John Romer served in Ireland then, from 1715, he was engineer at Sheerness, Tilbury, Gravesend and Portsmouth before being posted to Scotland, where he left a considerable legacy.

As at Inverness, the location of the second fort was of strategic importance. Wade proposed that a new fort – Fort Augustus (named in honour of George II's third son, William Augustus, later Duke of Cumberland) – should be built at the west end of Loch Ness, near Kiliwhimen (fig. 4.6). Fort Augustus was to be a 'modern fortification' (fig. 4.7), designed to awe the Highlanders and serve as the headquarters of the Hanoverian army. The governor of Fort Augustus was to be Commander in Chief of all the Highland forts, including Fort William and Fort George at Inverness, 'where a Body of 1000 Men may be drawn together from those Garrisons in 24 hours to Suppress any Insurrection of the Highlands'. Wade also recognised that Loch Ness formed a sensible means of transporting military provisions and troops between Fort George at Inverness and Fort Augustus and, on first coming to the Highlands, he had ordered an 'Exact Survey of the Several Lakes, Rivers, and Roads, between Fort William and Inverness' to be made (fig. 4.8).

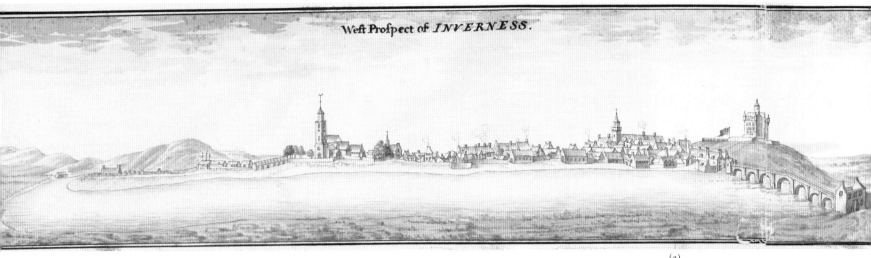

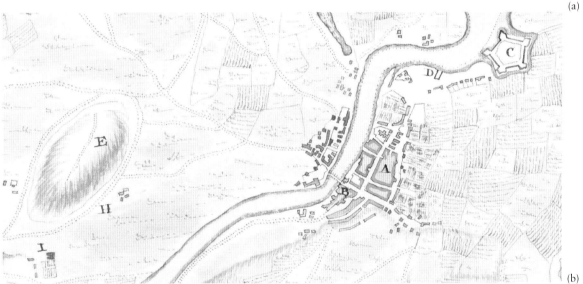

FIGURE 4.4

(a) When he was overseer of Bernera, John Henri Bastide assisted John Romer in surveying Inverness and Fort William. This view of Inverness from the west clearly shows the tower house of the medieval castle overlooking the seven-arched bridge across the River Ness.

(b) The view shown in (a) formed a small inset on a map of the country around Inverness, allowing the town to be visualised from both perspectives. The map is coloured according to established convention: red for masonry and buildings; sea-green (verdigris) for the outline of water courses, infilled with a paler wash; different shades of green to distinguish between marshy ground, gardens and arable fields; and hill-shading, at least in outline, in Indian ink. Bastide has used stylised symbols for areas of woodland, parallel hatching for cultivated land and parallel dotted lines for roads and tracks. The overall effect is an attractive imitation of nature but, as this extract shows, the focus of the map remains on features of military interest: the town (A), the castle (B), the old fort built by Oliver Cromwell (C), the pier (D), a nearby hill (Fairy Hill, E) and the 'Camp before marching to Glen Shields' (H).

Source: John Henri Bastide, *A General Survey of Inverness, & the Country adjacent to the foot of Loch-Ness; West propect of Inverness* (1725).

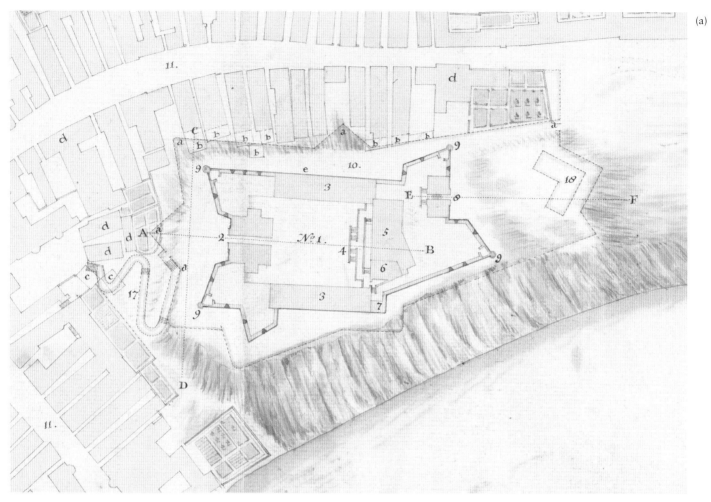

FIGURE 4.5

(a) (*Above*) Fort George in Inverness was founded on 7 August 1727. The fort was named after George II, who had succeeded to the throne that same year. John Romer, when first reviewing the old medieval castle at Inverness in 1725 with the purpose of converting it into barracks, found 'the charge will not answer the design, and at the best can be but a Crazy repair not much to be depended upon'. He proposed, instead, three 'Projects (in the nature of a Citadel) according to the situation'. His final design made use of the original five-storey tower house (marked as 5 on this plan), which he extended to the west and used as quarters for the officers, barracks for the gunners, and a storehouse (marked 6). The rest of the fort was new and included a three-storey soldiers' barrack (marked 3), a two-storey house for the governor (marked 2), and a powder magazine above which was a chapel (marked 8).

(b) (*Opposite*) As the sections and elevations across the various lines marked on the plan show, the site was cramped, with barrack accommodation blocks doing double service as ramparts, which were difficult to defend. The fort rapidly submitted to the Jacobites in February 1746.

Source: John Lambertus Romer, *A Plan of Fort George, & Part of the Town of Inverness, with Proper Sections Relating to the Fort* (1732).

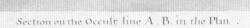

Section on the Occult line A, B, in the Plan.

Elevation of the Main Enterance of Fort George, through the line C, D.

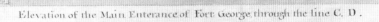

Elevation of the Sally port, and of the Chappel over it.

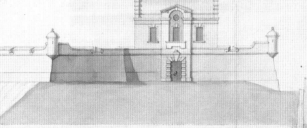

Section of the Sally port, and Chappel, on the line E, F.

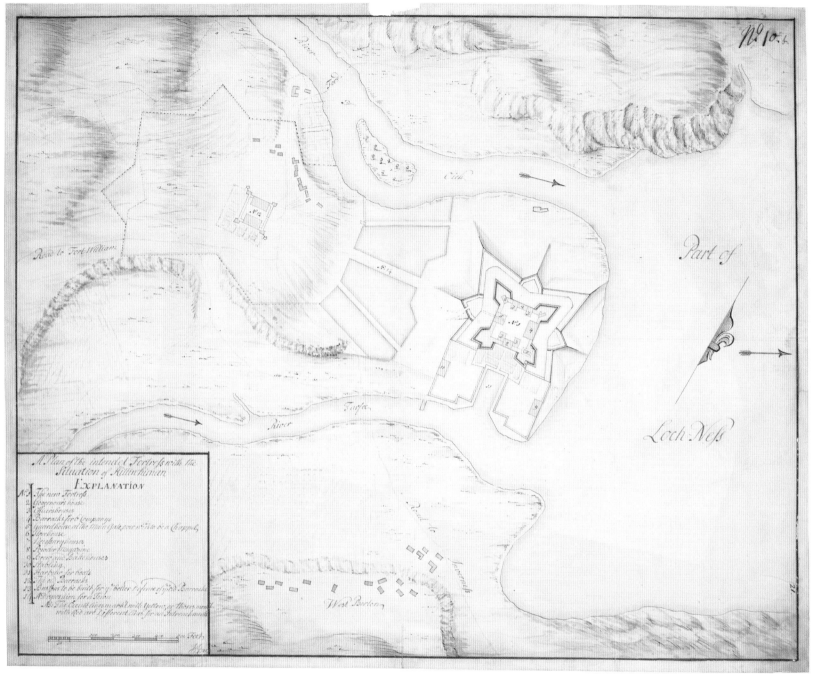

FIGURE 4.6

Wade had briefly considered upgrading the barracks at Kiliwhimen, but rejected the idea as the barracks were situated too far from the loch side to be readily supported in the event of a siege. At a scale of 200 feet to an inch (a prescribed 'Fortification Scale' encouraged by the Board of Ordnance for surveys of a fort or settlement), this plan of the environs of Kiliwhimen shows the position of the barracks (on higher ground to the left) and the location of the new fort, as well as roads, fords and planned entrenchments.

Source: John Lambertus Romer (and George Wade), *A Plan of the intended Fortress with the Situation of Killiwhimen* (1729).

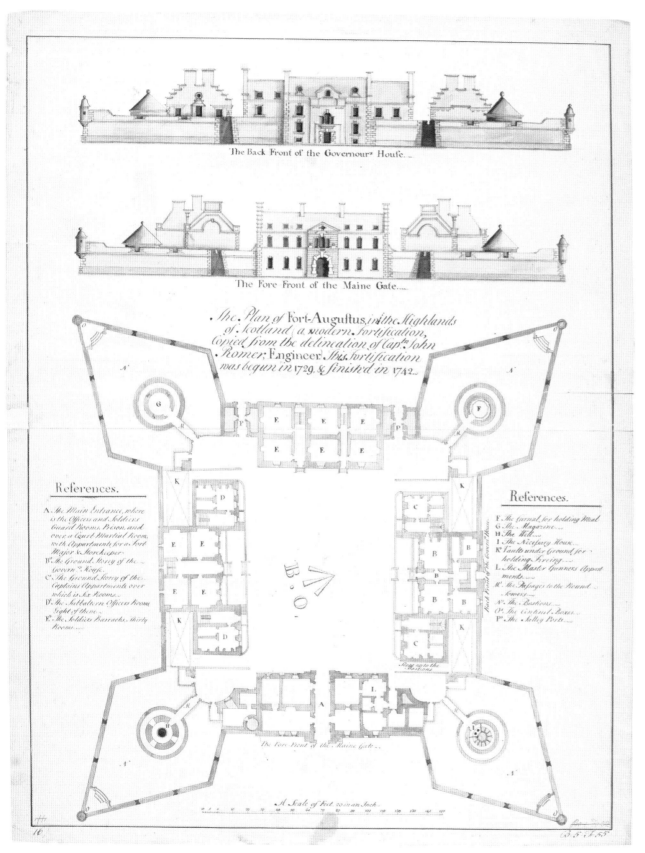
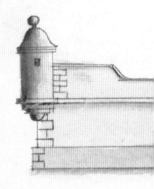

FIGURE 4.7

As a 'modern fortification', John Romer's design for Fort Augustus was ambitious, even if flawed. Construction work began in August 1729 and was eventually completed in 1742. The fort was regular in shape, comprising a square curtain wall with four massive stone-faced bastions at each corner, terminating in domed sentinel boxes – Romer's hallmark (see detail). Bordering the parade ground and rising above the curtain were four building blocks incorporating the entrance, the soldiers' barracks, the governor's house and the master gunner's apartments. 'Designed . . . more for ornament than strength', it had relatively thin curtain walls unable to withstand artillery, and was badly sited. Romer's detailed plan clearly shows the four pavilions inside each bastion. One was used for storing grain, the others as a well, a latrine and a powder magazine. The last proved to be the fort's undoing. In the 'Forty-five, Fort Augustus came under siege from the Jacobites in March 1746. The fort withstood fire for two days before a shell fired from Kiliwhimen half a mile away hit and detonated the exposed powder magazine, breaching the bastion walls, and the governor quickly surrendered.

Source: John Lambertus Romer, *The Plan of Fort Augustus, in the Highlands of Scotland (a modern fortification . . .)* (1742).

(a)

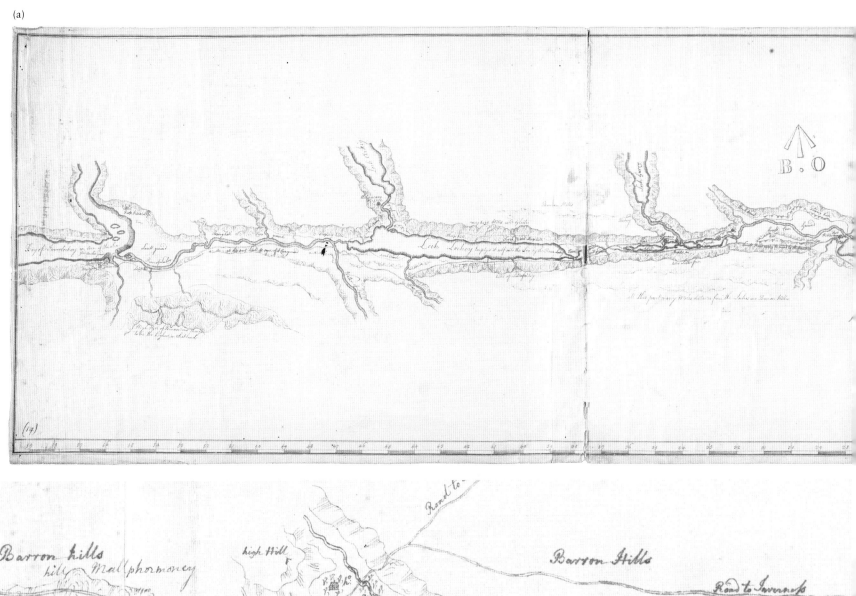

(b)

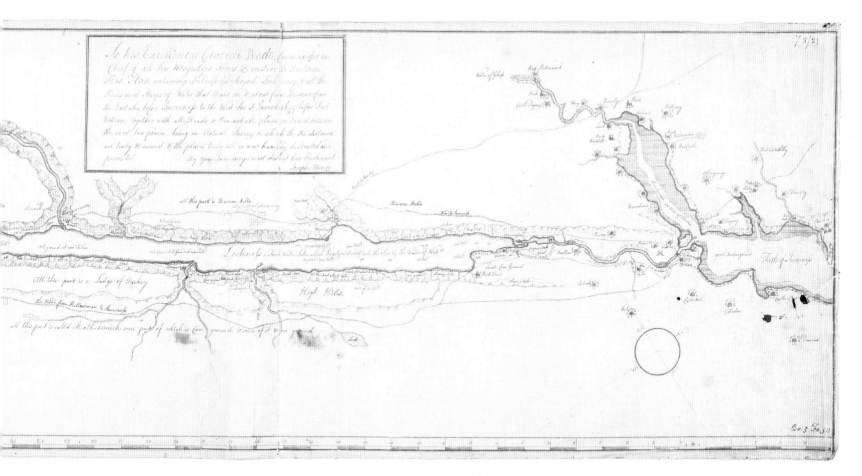

FIGURE 4.8

(a) (*Above*) Joseph Avery's survey of the Great Glen lochs is one of the earliest maps to show them correctly following a straight line. According to Avery, he completed 'an actual survey in which the distances are truly measured' and marked 'all Roads & Remarkable places' along the glen, setting it apart from any previous map of this area. The road from Fort William to Kiliwhimen, on the left, was the first section of the 'great Road of Communication extending from the East to the West Sea' to be built by independent Highland companies. As Avery was concerned to show waterways and routeways capable of transporting provisions, artillery and men, any detail beyond these features, such as the hills, was simply artistic in-filling.

(b) (*Left*) As this detail shows, Avery's map is also one of the earliest maps of Scottish freshwater lochs showing depths. Avery found that Loch Ness was 'Navigable for the Largest Vessells', the survey showing it to be '24 Miles in Length, and a Mile or more in breadth, the Country being Mountainous on both sides'. George Wade commissioned a small sailing vessel to be built 'sufficient to carry a Party of 60 or 80 Soldiers' and, in January 1726, the ship – the *Highland Galley* – was launched and proceeded to transport provisions and ammunition between garrisons at Inverness and Fort William.

Source: Joseph Avery, *This Plan Containing Lochness, Lochoyoch, Lochlochey, & all the Rivers and Strips of Water . . . from the East Sea before Inverness to the West Sea at Inverlochey . . .* (1727).

The focus was not purely on new fortifications. On first arriving in Edinburgh, Wade discovered that the castle was badly neglected and vulnerable to the Jacobite threat, as nothing had been done to improve its defences since 1715. On further inspection, he found the parapet walls 'so ruinous that the Soldiers after Shutting of the Gates had found ways to ascend and descend to, and from the Town of Edinburgh'. When ordered to try to scale the rock face and parapet, four armed soldiers managed it in less than five minutes! Attention turned to the defences at the west end of the castle, to the curved wall which made it impossible to defend from inside the fort (fig. 4.9). Similar defensive works were carried out at Dumbarton, along with works to expand barrack accommodation at Edinburgh, Stirling, Dumbarton and Fort William.

In his 1724 report, Wade noted that regular troops were at a disadvantage in the mountainous terrain of the Highlands, made 'still more impracticable, from the want of Roads [and] Bridges', but he did not explicitly propose constructing new roads until April 1725, when he sent the king a revised scheme. In this, he stated provision was needed 'for mending the Roads between the Garrisons and Barracks for the better Communication of His Majesty's Troops'. The following year he suggested that an annual sum of money was needed for constructing the roads and providing a salary for a dedicated inspector of roads. By 1730, Wade was granted an annual budget of £3,000.

A road system already existed in parts of Scotland before the eighteenth century, but many of the routes were passable only by men on horseback or foot, or they were used for droving animals. Few roads went through the central and western Highlands. Under Wade's direction, independent Highland companies set about 'opening ... great Road[s], by means of which, the Guns, Carriages &c can move without the least possibility of delay, in every direction'. Between 1726 and 1737, they constructed 259 miles of new roads and forty bridges. Although Wade is chiefly remembered today for the network of military roads and bridges in Scotland, many roads were, in fact, only planned by Wade, and some were never built (fig. 4.10). It was left to his successor, Major William Caulfeild, to manage the majority of road construction in the 1740s and 1750s (see chapter 5).

The first road to be built under Wade's direction ran from Fort William to Kiliwhimen. Building began in 1725 and, despite difficulties resulting from rocks, bogs and mountains, it was extended along the south shore of Loch Ness to Inverness by 1727, thereby connecting the garrisons at Fort William, Kiliwhimen and Inverness (see fig. 4.8a). Wade himself had travelled along the road to Fort William the previous summer 'in a Coach and Six Horses, to the great wonder of the Inhabitants' and reported that it was now fit for artillery vehicles and wheeled carriages.

Attention then turned to the central Highlands and the construction of two major north–south routes through the mountains (fig. 4.11). The first ran from Dunkeld to Inverness, a distance of 102 miles, and was completed in 1729; the second ran from Crieff to Dalnacardoch, where it joined the Dunkeld to Inverness route. This road, totalling 43 miles in length, was started and completed in 1730. In 1731, the Dalwhinnie to Fort Augustus road, by way of the Corrieyairack Pass, was completed.

In 1740 supervision of road construction devolved to Major Caulfeild. Coinciding with the change of personnel were heightened fears over a Jacobite rising, which prompted copies of maps of Wade's network of military roads to be published (fig. 4.12). Military commanders on both sides seem to have found these maps useful. Wade's roads were built for strategic reasons, for improving the speed and ease of troop movements between military establishments to counter any insurrection. The Jacobites certainly recognised this. In September 1745, it was reported that 'some of [the rebellious MacDonald] Chiefs stay'd at home [and] were breaking down the Bridges & ruining the Roads' to hinder the Hanoverian army. Ironically, as the 1745 Rising progressed, the benefit that the roads afforded the Jacobite army caused government troops to destroy their own constructions: intelligence from London disclosed 'an Order from Lord Launsdale for breaking up the Roads &c. &c. doing every Thing [that] can retard the Rebels in their March'.

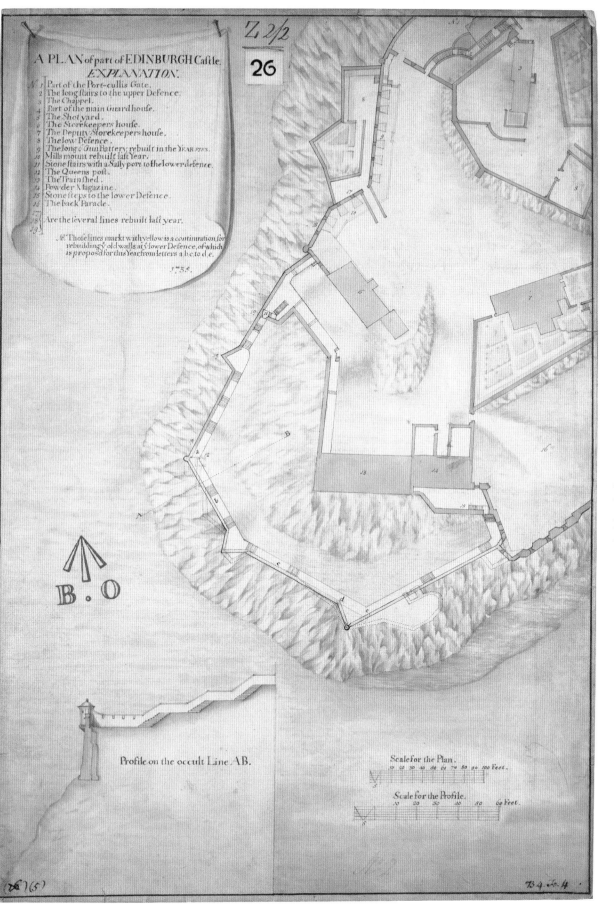

FIGURE 4.9

Under Wade's orders and to John Romer's design, Bucknal, the Fort Major, directed the building of a new, angled wall to better secure Edinburgh Castle. The design included a large buttress and a sentinel box with further indications of the work to be completed later that year, including the addition of barriers, the wall to be raised by 5 feet to secure the upper defences and, later, a new powder magazine. These defence measures and those to the castle entrance proved highly effective. The castle was one of the few Hanoverian fortifications not to fall during the 'Forty-five.

Source: Anon., *A Plan of Part of Edinburgh Castle* (1735). Courtesy of the National Records of Scotland.

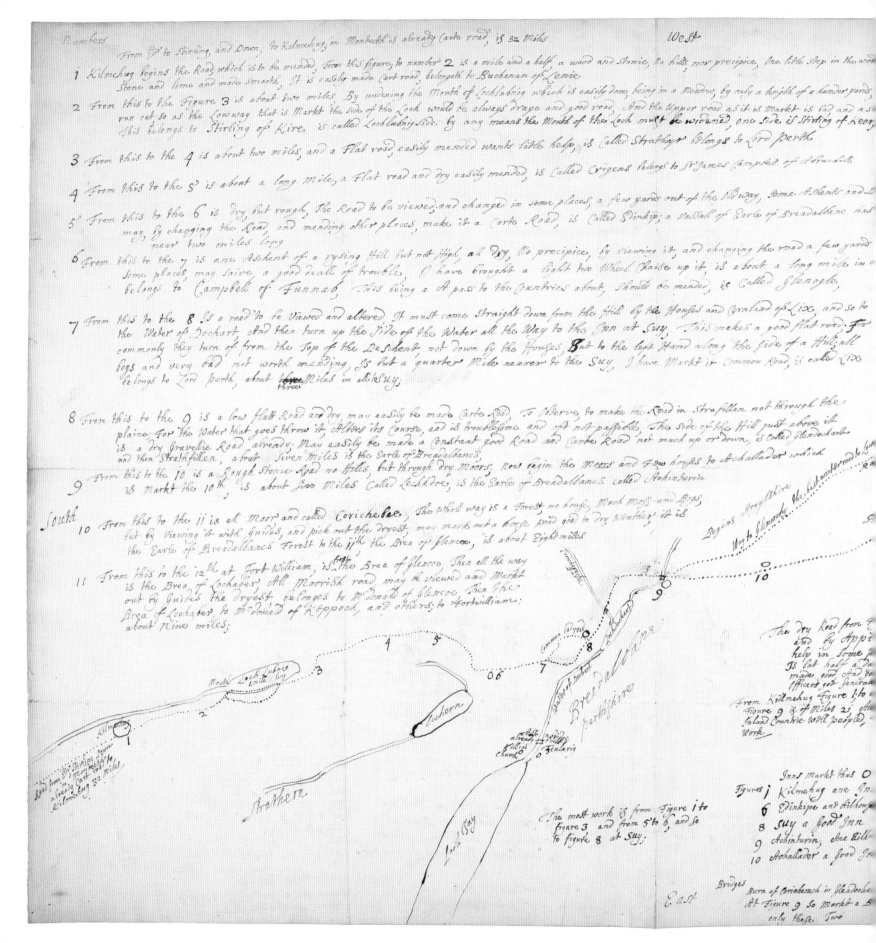

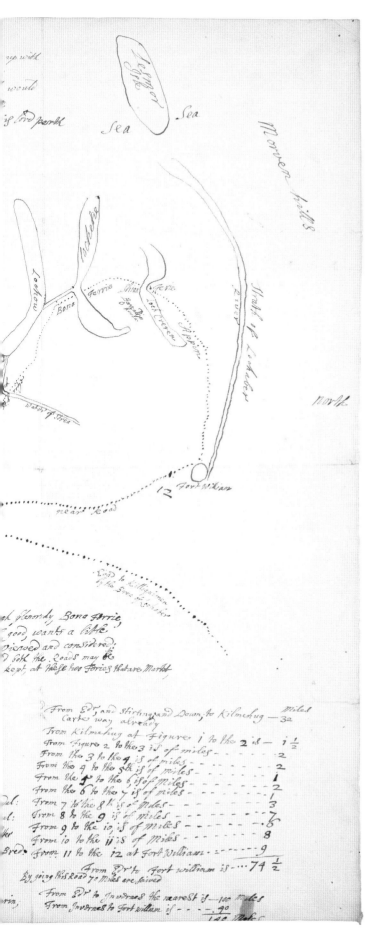

FIGURE 4.10

Wade often represented proposed routes in two forms, map and written itinerary, presented on the same sheet. His work offered a utilitarian description of the geography of parts of the Highlands; it was the content of the maps rather than their appearance that mattered most. This survey report describes in detail the condition of the proposed route between Callander and Loch Tay to Fort William and Appin. Wade's sketch begins at the inn at Kilmahog in Perthshire, 32 miles from Edinburgh by 'carte way' and, thereafter, is split into twelve sections of distances varying between 1 mile and 9 miles. In total, 42½ miles of track are described. He also indicates where the track needs minor attention from the road engineers – '4 From this to the 5 is about a long mile, a Flat road and dry easily mended' – or needs to be completely rebuilt – '7 from this to the 8 Is a road to be viewed and altered. It must come straight down from the Hill by the Houses and Cornland.' As well as providing information pertinent to road engineers, Wade's survey also noted whose land the road passed through. Since this identified whether the landowner was for or against government, it revealed the security of the road for use by military troops. '11 From this to the 12th at Fort William', for example, 'is first the Brea of Glencoe, Then all the way is the Brea of Lochaber, . . . belonges to McDonald of Glencoe, Then the Brea of Lochaber to McDonald of Koppoch'. Both MacDonalds had joined the Jacobites in the 1715 Rising.

Source: George Wade, Sketch and description of the proposed roads from Callander and Loch Tay to Fort William and Appin including details of mileage and of inns (1724–45).

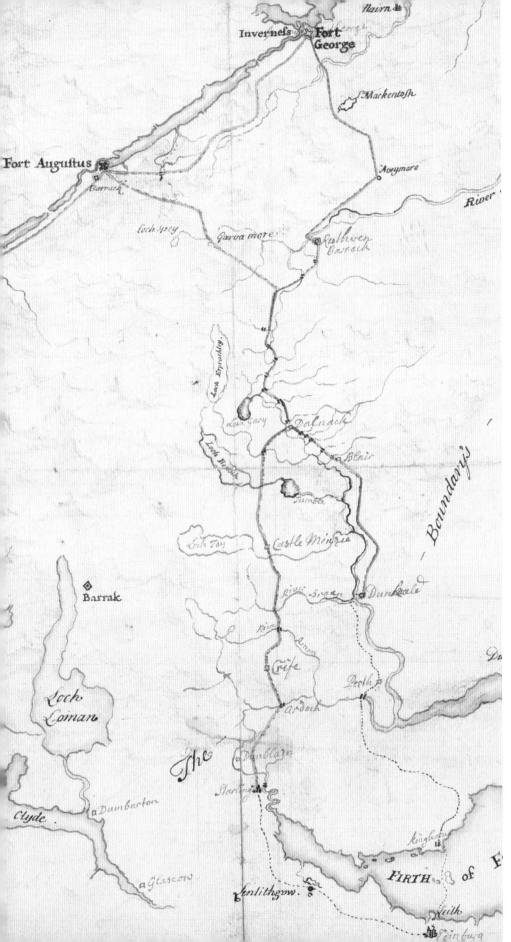

FIGURE 4.11

Wade intended that two north–south military routeways should be built, one from Inverness to Perth and the other to Stirling, to facilitate a 'Short and speedy Communication' with the troops quartered in the Lowland garrisons. While both roads were completed as far as Crieff and Dunkeld by 1730, it was left to Wade's successor, Major William Caulfeild, to complete the southern extents after 1740. Although the roads were still under construction in 1728, they were already making a noticeable difference to travel in the Highlands. In a letter to Lord Townsend dated 23 August 1728, Duncan Forbes, the Lord Advocate of Scotland, wrote:

> In coming from Perth I chose the Highland road, By Blair in Atholl, . . . and in my journey had great Relief . . . I was not a little surprised at the Regularity and success of the work . . . [the road is] now as Smooth as the Road from London to Hampstead, and in a little time will be passable by an Army with its Artillery, notwithstanding the abrupt Declivity of some of the Mountains. In short this Project which compleats the intention of Disarming the Highlands is so far from being ill Received . . . it is looked upon as a sort of satisfaction for the Loss of their Weapons, since it gives the Troops access to come wherever there is Occasion to Defend them, So well has Mr Wade known how to Guile the Pill that Deprives them of the power of hurting the Government.

Source: George Wade, Map showing the intended military roads joining up Stirling with Fort Augustus, etc. (1725).

FIGURE 4.12

(a) This is one of the most up-to-date maps showing roads in central Scotland that would have been available to military commanders in the 'Forty-five. In a letter dated December 1745, Lieutenant General Henry Hawley, on his way to take command of the Government troops in Scotland, wrote: 'I am going in the dark; for Marechal Wade won't let me have his map; he says that his majesty has the only one to fellow it. I could wish it was either copied or printed, or that his majesty could please lend it to me.' We are not sure which map Hawley alludes to; it could be Clement Lemprière's 1731 map *A Description of the Highlands of Scotland* (fig. 3.18), one of Wade's surveillance maps of Scotland or a road map showing the Highland routes such as this one published by Richard Cooper. What Hawley's letter does reveal is how little was known of the extent and geography of Scotland, especially the Highlands, at the time of the 'Forty-five, despite the activities of the Board of Ordnance engineers and surveyors. Cooper's lochs and rivers are largely taken from the Blaeu outlines of the previous century, overlaid with a very generalised road network, completely impractical for military manoeuvres or for confronting an army in hostile country.

Source: Richard Cooper, *A Map of His Majesty's Roads from Edinburgh to Inverness, Fort Augustus & Fort William . . .* (c.1742).

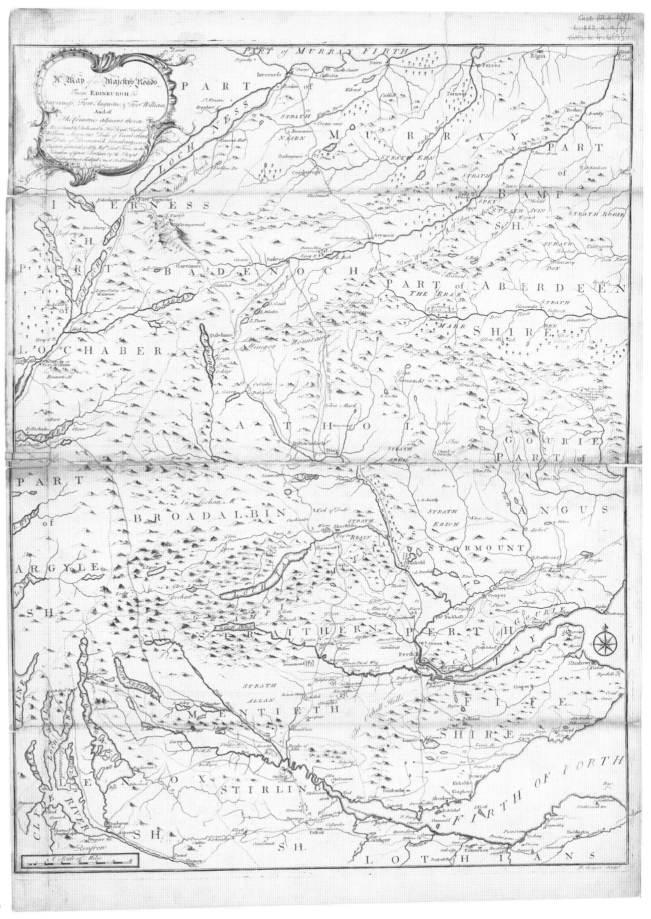

(a)

FIGURE 4.12 (*continued*)

(b) The detail shows Wade's road over the Monadhliath Mountains via the Corrieyairack Pass. Seventeen traverses were constructed to tackle the steep ascent, each one commanded by the one above, which gave anyone descending a distinct advantage. From the very start of the 'Forty-five, the Jacobite army made full use of Wade's military roads. In August 1745, Charles Edward's army marched south from Aberchalder at the north end of Loch Oich to seize the high ground of the Corrieyairack Pass, where they expected to engage the government army. But Lieutenant-General Sir John Cope, Commander in Chief of the British forces in Scotland, had turned his army around and headed for Ruthven in Badenoch instead. This left the pass open, and the Jacobite army marched southwards into Perthshire unchecked.

Source: Richard Cooper, *A Map of His Majesty's Roads from Edinburgh to Inverness, Fort Augustus & Fort William . . .* (*c*.1742).

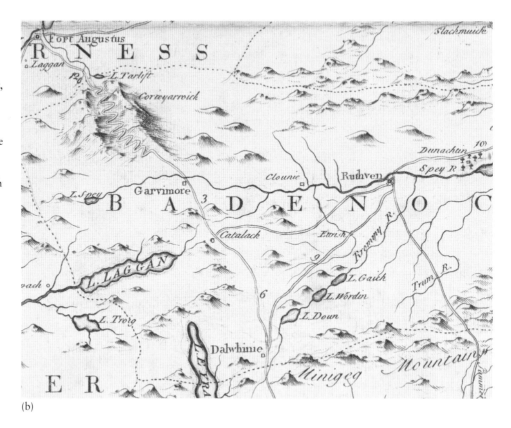

(b)

Tactics: The 'Forty-five

Despite a change in military practices in Scotland, Jacobitism continued to survive in several isolated but intensely committed local communities and they in turn successfully nurtured a network of communications between Scotland and western Europe. In 1744, a planned invasion of England by Louis XV of France was aborted when the weather conspired against a Jacobite-led assault. A year later, two ships – the *Du Teillay* and the *Elisabeth* – set sail on 4 July from Belle-Isle in France. Charles Edward Stuart, the Young Pretender, was on board the *Du Teillay*, while the *Elisabeth* carried 1,500 firearms, 800 broadswords, 20 small cannon and a contingent of about 100 men. Both ships were bound for Scotland, but only the *Du Teillay* made anchor (at Loch nan Uamh); the *Elisabeth* had been badly damaged during an encounter with the *Lion*, an English man-of-war, and was forced to return to France. From Loch nan Uamh, Charles Edward travelled to Glenfinnan at the head of Loch Shiel in Moidart and raised the Royal Standard on 19 August 1745.

Orders were sent to Lieutenant-General Sir John Cope, Commander in Chief of the British forces in Scotland, to 'assemble the troops in proper places'. He recalled the military parties at work upon the roads and sent orders for the various units to assemble at Stirling. From there, Cope mustered a field force and marched them via Crieff towards Fort Augustus. To general astonishment, at the last minute Cope turned away from engaging the Jacobite forces at the Corrieyairack Pass, instead diverting his course to march north to Inverness. This left the Lowlands open to Charles Edward and his supporters (see fig. 4.12). Cope later claimed that his motive had been to secure Inverness. He was given the benefit of the doubt, but, to observers at the time, his real fear of meeting the Jacobite army in hostile and unknown terrain seemed more important. The Jacobite army advanced by way of Wade's roads to seize Perth before advancing on Edinburgh where, on 18 September, they secured the city but not the castle (figs 4.13 and 4.14). Meanwhile, Cope marched his army from Inverness to Aberdeen and from there embarked by ship to land at Dunbar on the same day that Charles Edward gained Edinburgh.

On 21 September 1745, the two armies came face to face at Prestonpans (also known as Gladsmuir). Cope and his troops were surprised by a dawn advance by the Jacobites. The battle lasted no more than eight minutes and resulted in a resounding defeat for the Hanoverian army. Several contemporary maps illustrate events. Some focus on the changing deployment of the armies leading up to the point of engagement or the order of battle, while others are printed records of events – 'news' maps (fig. 4.15).

Why were so many maps made of an undisputed government defeat? It is possible that maps were intended as evidence for the public enquiry into Cope's handling of the 'Forty-five. We know that a map drawn by William Roy, Captain of Engineers and Assistant Quartermaster-General of the British Forces during the Seven Years' War (1756–63), was submitted as key evidence at the trial of Lord George Sackville after he failed to advance at the battle of Minden. Surprisingly, Cope was ultimately acquitted by a court-martial of any major misconduct in the campaign and battle. Maps of defeat were also useful in informing future military planning and battle tactics. There was great public interest in the battle, too, and how quickly the government's forces had been defeated. One publication offered the public 'an enquiry into the conduct of General Cope' and advised the reader 'now . . . to cast his Eye on the Plan of the Battle'.

Following their victory at Prestonpans, the Jacobite army marched into England as far as Derby. When the expected support from France and an English Jacobite rising failed to materialise, however, the decision was reluctantly taken to retreat to Scotland. On 5 January 1746, the Jacobites took the town of Stirling. Stirling Castle, having undergone almost continuous refortification since the Union, withstood their advance and remained in Hanoverian control under the command of Major General William Blakeney (fig. 4.16). In response to the Jacobite siege of the castle, a Hanoverian army commanded by Lieutenant-General Henry Hawley, new Commander in Chief of the forces in Scotland, marched north, stopping at Edinburgh to assemble more troops, before marching north-west towards Stirling. Charles Edward, intent on confronting Hawley before he reached Stirling, marched a Jacobite force to Plean Muir south-east of the town. Hawley's army, meanwhile, moved more slowly, and late in the afternoon of 16 January they encamped on low-lying ground between Falkirk and Carron Water to the north.

The battle of Falkirk was fought on 17 January 1746 (fig. 4.17). The terrain on which the battle was fought was particularly difficult and was used to great advantage by the Jacobites, who took the initiative, leaving Plean Muir in the early hours of 17 January to march to and deploy on commanding ground to the south-west of Falkirk (Falkirk Muir), an uneven rolling plateau that dominated Hawley's encampment. The Hanoverian army had a steep climb into a strong south-westerly wind and driving rain to reach the battlefield. From this time, the surviving maps give no visual indication of the action that took place. As it was, the Hanoverian Dragoon charge failed after a violent encounter with the Highlanders: 'The resistance of the Highlanders was so incredibly obstinate, that the English, having been for some time engaged pell-mell with them in their ranks, were at length repulsed and forced to retire.' The Hanoverian right wing, however, held their line against the Jacobites, protected by the ravine, and in the intense barrage of musket fire, some Jacobites did take flight. Failing light and bad weather discouraged Hawley from seeing this small advantage through to the end and, rather than going on the offensive, he retreated to his camp at Falkirk, then to Linlithgow. The Jacobites claimed a victory and at least one map survives as an important piece of propaganda (fig. 4.18).

News that the Duke of Cumberland was expected in Edinburgh with three more regiments spread a mood of despondency among the Jacobites at the prospect of another

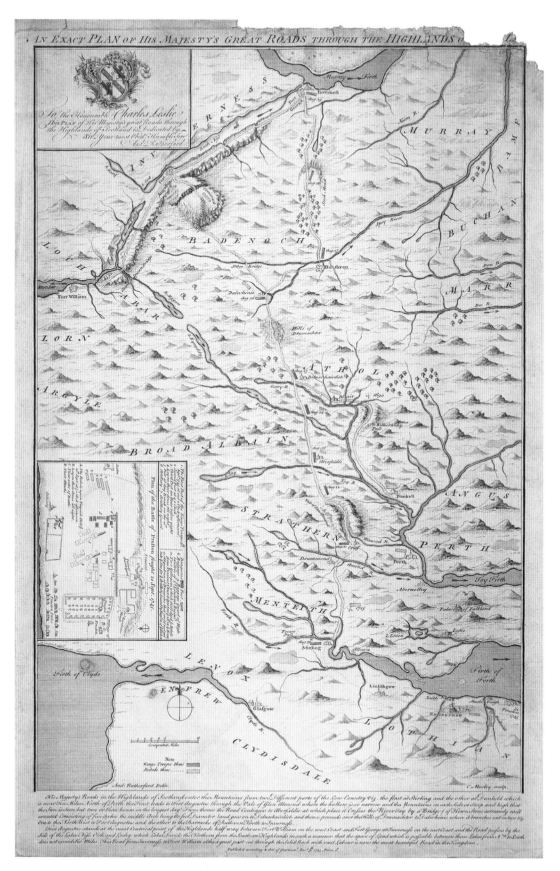

FIGURE 4.13

Andrew Rutherford's printed map, for a London audience eager for detailed information on the Jacobite army and their initial successes in the 'Forty-five, shows the respective marches and encampments of the Hanoverian and Jacobite armies in August and September 1745. A short description below the map provides details of the network of military roads connecting the Lowlands with the government forts and garrisons in the Highlands. Cope marched north from Stirling on the same day the Royal Stuart Standard was raised on 19 August, reaching Crieff the following day, Aberfeldy on 23 August and Dalwhinnie on 26 August. At Dalwhinnie, he was faced with negotiating the Corrieyairack Pass, where he was in danger of being ambushed by the advancing Jacobite army. Instead, he diverted to the barrack fort of Ruthven in Badenoch to the east and from there marched north to Inverness, arriving on 29 August. Meanwhile, the Jacobite army took advantage of the open road south and reached Blair Castle on 31 August, where they rested for two days before advancing to Dunkeld. On 4 September, they reached Perth. By 15 September, they had crossed the Fords of Frew (fig. 4.14) west of Stirling and reached Linlithgow before securing the town of Edinburgh on 18 September (see detail). In the north, Cope and his troops marched from Inverness to Aberdeen from where they sailed south, disembarking at Dunbar on 19 September and marching west towards Edinburgh. On 21 September, the two armies finally met near Prestonpans.

Source: Andrew Rutherford, *An Exact Plan of His Majesty's Great Roads through the Highlands of Scotland* (1745).

(a)

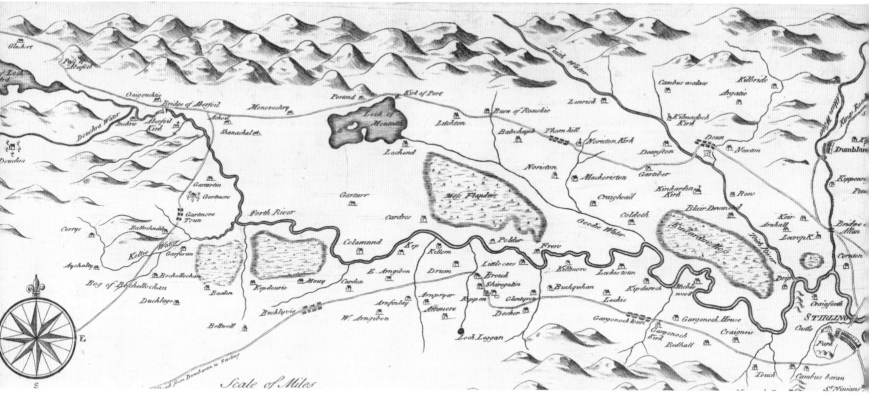

FIGURE 4.14

William Edgar's map of the River Forth west of Stirling includes the point at which the Jacobite army crossed the river, at the Fords of Frew (shown bottom centre on detail (b)), as they marched south (13 September 1745) and then again on their way north (1 February 1746). The *Description* below the map describes the ford as very shallow in the dry season, but a small amount of rain makes it impassable. Another version of the map also exists, but with the *Description* in manuscript (in the British Library).

A comparison of the two maps reveals differences between the manuscript and printed texts. The manuscript text describing Doune Castle ends 'an old strong place belonging to the Earl of Murray, presently possest by the Rebels'. Doune Castle ('Doun', to the top right of detail (b)), the seat of the Earl of Moray, was taken by the Jacobites in 1745. The map with printed text lacks this final reference to the rebels, suggesting the manuscript text version dates from during the 1745 Rising and the printed version is later. An engraving of the map was commissioned by Lord Justice Clerk, Andrew Fletcher, Lord Milton, who may have sent the version with manuscript text to the Duke of Cumberland. This map was especially valuable to the Hanoverian forces in providing a very detailed topography of a key strategic area during the 'Forty-five.

Source: William Edgar, *Description of the River Forth above Stirling* (1746).

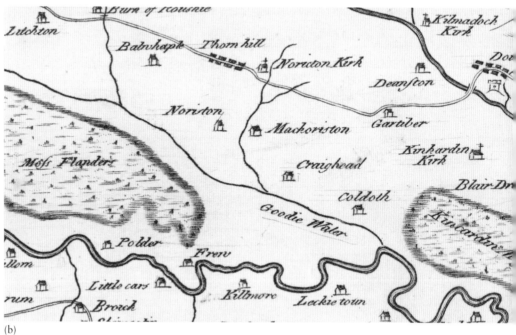

(b)

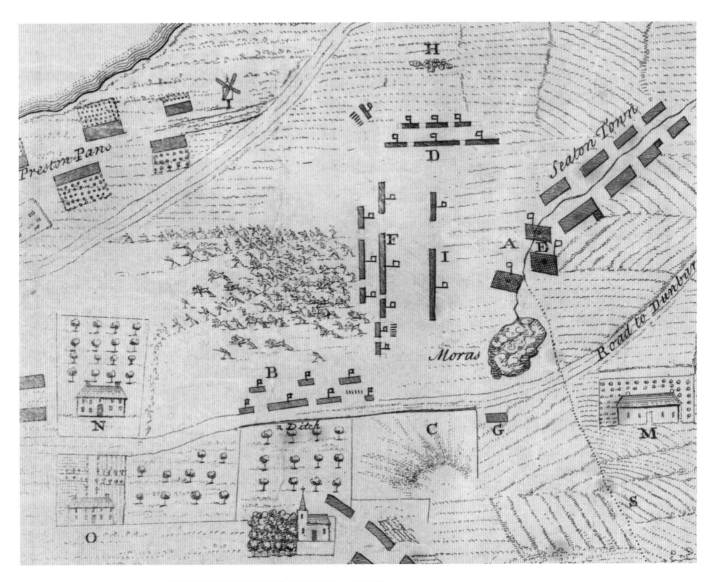

FIGURE 4.15

This 'news' map shows the Jacobite rout of the Hanoverian army under Cope at the battle of Prestonpans. The draughtsman – 'an Officer of the Army who was present' – has managed to portray the dynamics of the brief moments of action using letters explained in a 'Table of References'. These show how Cope drew up his army to face the Jacobites to the south, and how the Jacobites then cleverly circled around them to the east to form at the rear of the Hanoverian army, thus taking Cope by surprise. The map was published on 6 November 1745 and sold by J. Collyer in Ludgate Street and G. Woodfall near Charing Cross for one shilling 'plain'. News maps provided the most complete illustration of events with claims to be eyewitness accounts, as shown here.

The TABLE of REFERENCES.

A THE Place where Sir *John Cope* drew up his Army, at his first coming to the Field, 20 *September*.
B The Form of his Army, when they expected the *Highlanders* would march down the Hill C, and attack them; in Expectation of which they had moved from D, and to which Ground they returned about 3 o'Clock in the Afternoon, and there remained until the *Highlanders* appeared, about 5 next Morning, in three Columns E, when they left that Ground, and formed at F.
G Sir *John*'s advanced Guard.
H His Baggage.
I The *Highland* Army when formed.

Source: Anon., *A Plan of the Battle of Preston Panns fought 21st Sept. 1745* (1745).

FIGURE 4.16

(*Above*) This perspective view of Stirling Castle by John Elphinstone shows the castle under siege from the Jacobite army. Without heavy artillery, the Jacobites were unlikely to take the castle, but their main aim was to draw General Hawley and his army away from Edinburgh Castle to a battle ground of their choosing. When reports of a large body of troops moving westwards reached the Jacobites, they retreated to Falkirk and then to Bannockburn. Following the battle of Falkirk (see figs 4.17 and 4.18), Stirling Castle's commander, Major General Blakeney, opened fire and demolished the Jacobites' battery, killing many of the French and Lowland troops manning it.

Source: John Elphinstone, *A New Map of North Britain* . . . (1746). Courtesy of the British Library Board.

FIGURE 4.17

(*Opposite*) This map illustrates events from the evening of the 16 January 1746 to the commencement of the battle of Falkirk the following day (south is to the top of the map). Another almost identical map exists. The only notable difference between them is the inclusion (on this map) of a title cartouche and a dedication to the Duke of Cumberland. It is therefore likely that this map was drawn at a later date (a fair copy). As Falkirk proved to be another unmitigated defeat for the Hanoverian army, it is less a commemoration than a record of the army's tactical manoeuvres and sequence of troop deployments in relation to the topography. A version of his map was published in *The London Magazine, and Monthly Chronologer* (1746) to accompany a narrative of events.

Source: William Cunningham, Map of the battle of Falkirk, 17 January 1746 (1746). The Royal Collection © 2018, Her Majesty Queen Elizabeth II.

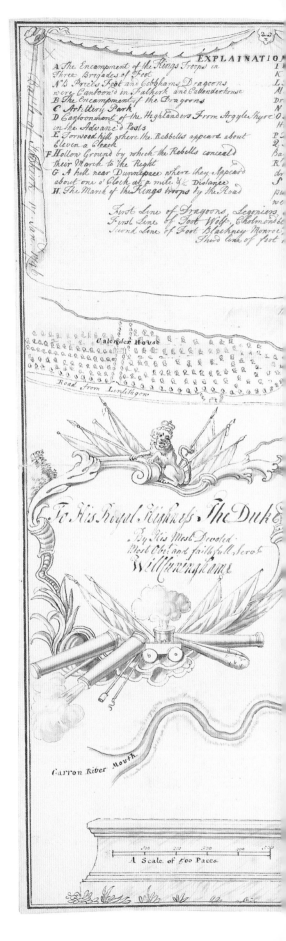

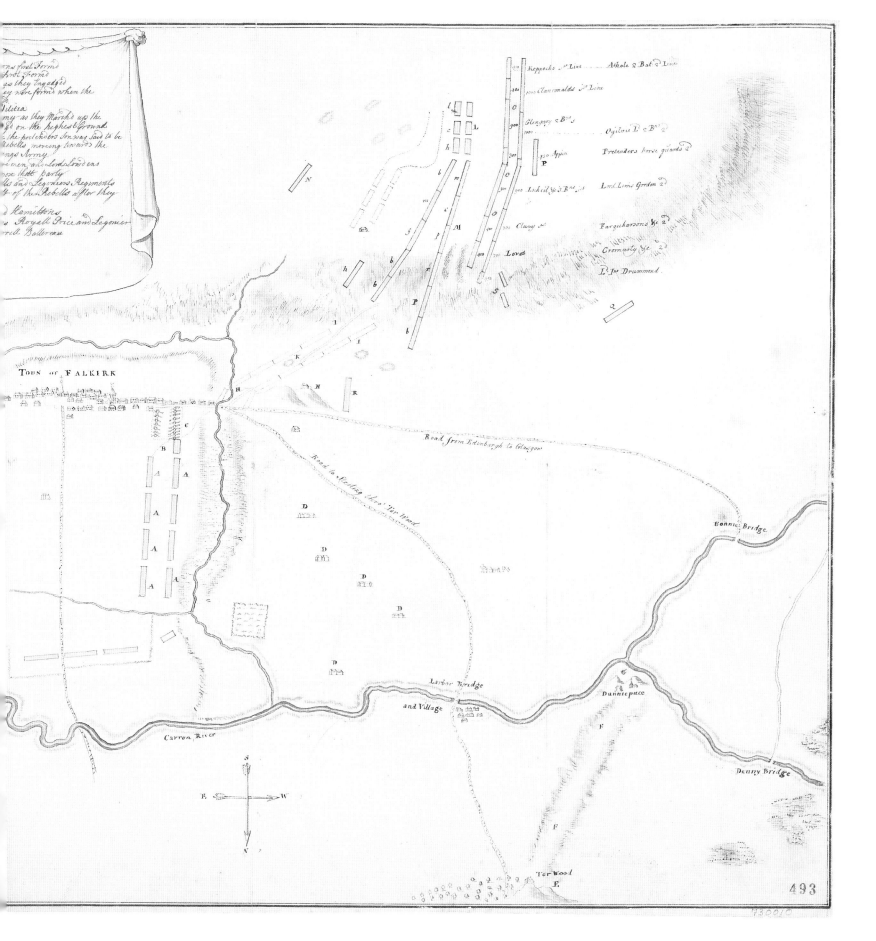

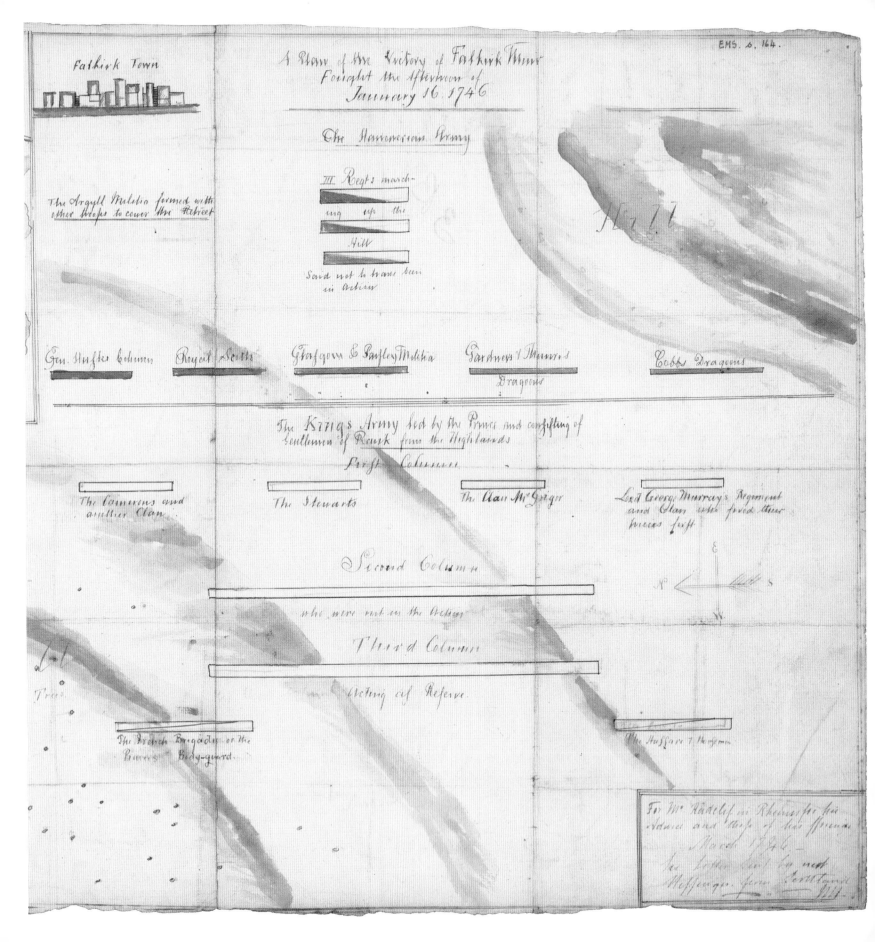

battle. Following a Council of War at Crieff, Charles Edward and his Jacobite army began to move north into the Highlands, heading for Inverness. On 10 February 1746, the Jacobite army captured Ruthven Barracks. Fort George at Inverness fell soon after, as did Fort Augustus. Only the garrison at Fort William successfully resisted the Jacobite siege (see fig. 3.8b). Meanwhile, Cumberland marched north towards Aberdeen (fig. 4.19).

The final battle of the 'Forty-five was fought on Culloden Moor on 16 April 1746. The Jacobite army, outnumbered, exhausted and outmanoeuvred, was no match for Cumberland's Hanoverian army. One of the best documented battles, Culloden is also one of the best mapped, including plans of battle order (fig. 4.20), sketches made immediately after the battle and post-battle records of the event. When Cumberland's army encamped at Inverness in the wake of the battle of Culloden, Joseph Yorke, Cumberland's *aide-de-camp*, noted in his Orderly Book an instruction that 'All Paper, Letter, Commissions, Maps or Plans taken in the Field of Battle or since, to be brought to H.R.H. Qrs. & delivered to Sr. Everard Fawkener', Cumberland's private secretary. Many of the surviving maps now form part of the Cumberland Collection at Windsor. However, two contemporary maps held at the National Library of Scotland offer interesting and slightly different perspectives on the battle. The first was compiled by a 'Lieut^r. Fireworker in ye Roy^l. Train of Artillery' (fig. 4.21); the second by an unknown French officer who may have been in the Royal Ecossais or Irish Picquets since the map shows more detail of the centre rear of the Jacobite army than any other (fig. 4.22).

The military maps during this period reveal new, imaginative, but still faltering attempts to deal with the significant Jacobite threat in Scotland. Lack of funds and manpower from the Board of Ordnance meant that new forts often had critical limitations, their reach was limited, while the road network to connect them was basic and, at times, of greater use to the Jacobites. Military mapping was localised to particular places, with no comprehensive survey or strategic overview of the whole country, a problem that was critically revealed to Hanoverian military commanders in the 'Forty-five. We examine the responses to this in the next chapter.

FIGURE 4.18

(*Opposite*) This plan of the battle of Falkirk is essentially an order of battle showing the front line of the Hanoverian army and the 'Kings Army led by the Prince and consisting of Gentlemen of Rank from the Highlands'. Importantly, the regimental lines are depicted in relation to the topography: a plateau – to one side bound by a morass and the other by a steep-sided ravine – which proved a strategic advantage to the Jacobites and a tactical disadvantage for Hawley's army. The Hanoverian army, however, did not see Falkirk as a defeat. According to Hawley, they were 'masters of the field of battle', which may in part explain why Cunningham's map (fig. 4.17) failed to describe events during and after the battle.

Source: J. Millan, *A Plan of the Victory of Falkirk Muir Fought the Afternoon of January 16 1746* . . . (1746).

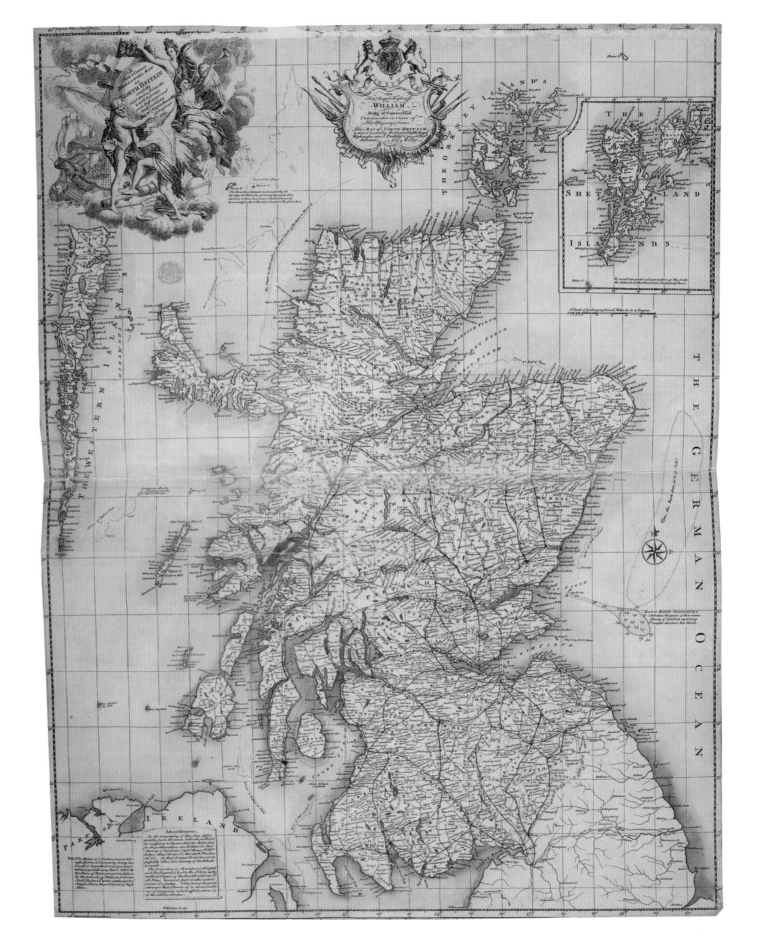

FIGURE 4.19

(*Left*) On leaving Crieff, the Jacobite army made their way north, pursued by William, Duke of Cumberland, as far as Aberdeen. Acutely aware of his lack of topographical knowledge and the difficulty of moving troops through the Highlands, Cumberland remained near Aberdeen for two months, gathering information about the roads and routeways of the region and the passage of the glens. The most up-to-date map of Scotland available to him at the time was this map by John Elphinstone, a military engineer in his army. Although the topographic content of the map is an improvement on earlier maps, it still reflects the outline and form of Scotland depicted a century earlier in Blaeu's *Atlas Novus* (1654), such as the bend in the Great Glen and the shape of Skye. More problematic for military purposes, the depiction of the hills was entirely schematic, without the detail necessary to accurately plan routes or military offensives. This was not lost on Elphinstone, who notes:

> As the Geography of this Map differs greatly from all others hitherto publish'd; it's necessary to observe that the authorities for these alterations are Mr. Adair Sr. Alexander Murray of Stanhope Captn Bruce William Edgar Alexander Bryce & Murdoch Mackenzie &c. So that it must be as Correct as possible till a New Survey of the Whole is made.

This copy has been annotated by David Watson, Deputy Quartermaster-General of the Ordnance, to show the roads and 'posts proposed to be occupied by the regular troops in the Highlands', most likely as part of the military actions after the 'Forty-five (see chapter 5).

Source: John Elphinstone and David Watson, *A New & Correct Mercator's Map of North Britain* (1746). Courtesy of the British Library Board.

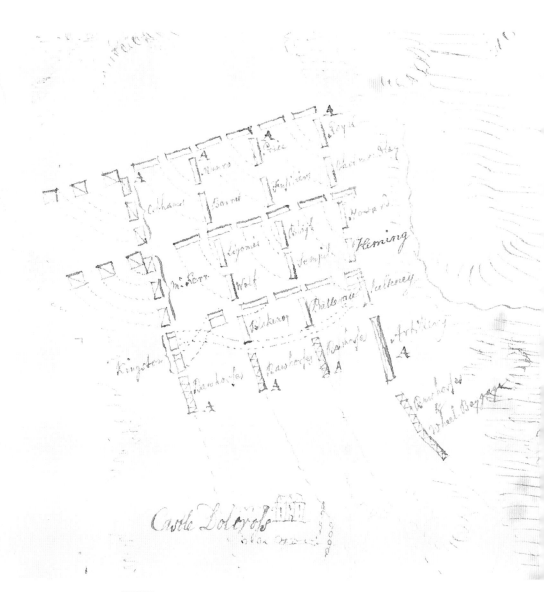

FIGURE 4.20

By the time Cumberland reached the battlefield at Culloden, he had perfected the Hanoverian army's order of march to the extent that it could 'swing' straight into the predetermined battle formation: 'we marched in four columns, and by the ruff of a Drum formed instantly in to order of Battle'. [Henry] Schultz, Cumberland's personal draughtsman, uses rectangular symbols to distinguish the different units and broken lines to indicate their manoeuvres.

Source: [Henry] Schultz, *Plan of the Battle of Collodden* (1746). The Royal Collection © 2018, Her Majesty Queen Elizabeth II.

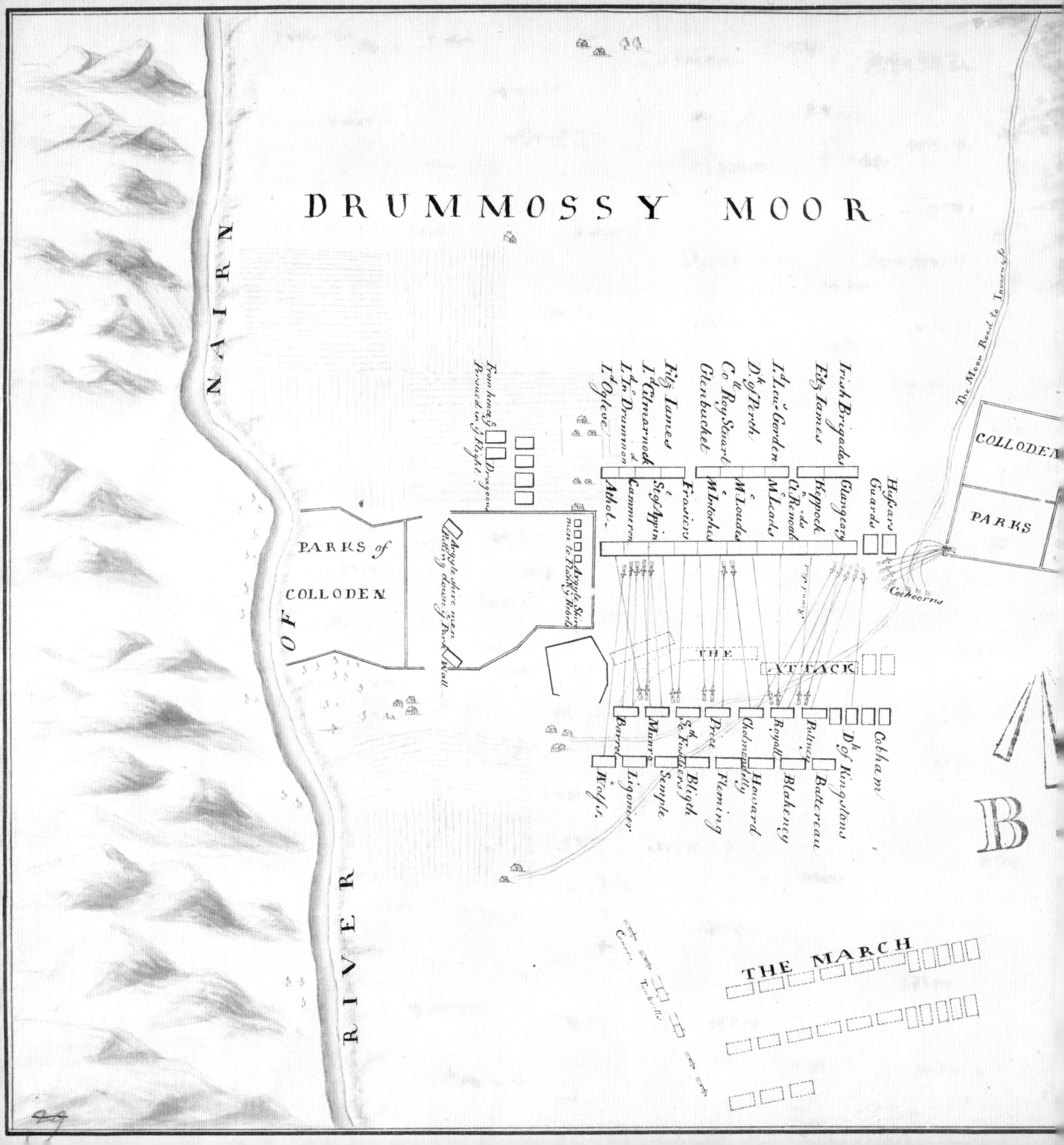

DRUMMOSSY MOOR

NAIRN

RIVER OF

PARKS of COLLODEN

COLLODEN PARKS

The Moor Road to Inverness

Argyle Shire men to Parks Wall
From hence ye Dragoons pursued in ye Flight

Hybers Guards
Irish Brigade
Fitz James
Glangery
Keppoch
Ct. Renald
Ld. Lui. Gordon
Dk. of Perth
M. Leods
Co. Roy Stuart
Glenbucket
M. Intochs
Fitz James
Fraziers
Ld. Kilmarnock
St. of Appin
Ld. Drummond
Camerons
Ld. Oglevie
Athol

Cochoorns

THE ATTACK

Dk. of Kingstons
Cotham
Pulney
Battereau
Royall
Blakeney
Cholmondly
Howard
Price
Fleming
St. Fusileers
Bligth
Munro
Semple
Barrel
Ligonier
Wolfs

THE MARCH

B

FIGURE 4.21

This battle plan of Culloden by the Hanoverian artillery lieutenant Jasper Leigh Jones is most useful for showing the disposition of the artillery. He signifies the artillery train to the left of the main military march with labels for 'Cannon' and 'Tumbrells' (two-wheeled carts) and his depiction of the action is focused on the artillery, with little attention paid to the dragoons and their breach of the Culwhiniac enclosures. On the Hanoverian side, he shows five groups of two cannons positioned in advance of the front line. Four further cannons and three Coehorn mortars (lightweight mortars, originally designed by the Dutch military engineer Menno van Coehorn (1641–1704)) are depicted in a forward position to the south-east of Culloden Park. These cannon and Coehorns are, in fact, 'time-lapsed' depictions, having been brought up from the Hanoverian front line during the battle. The Jacobite army has two centrally positioned cannons flanked on either side by groups of five cannons. Between the armies' front lines, Jones marks the advance of the Hanoverian army as 'THE ATTACK' and shows, with apparent accuracy, the trajectory of the cannons and mortars.

The English traveller, Richard Pocock, who recorded details of the battle in his *Tours in Scotland 1747, 1750, 1760*, described some of the fiercest action in more detail:

> Our forces to the left were drawn up on a rising ground much lower than theirs [the Jacobites] [and] stretching beyond their right line with a small shallow valley and a bed of a winter stream between them . . . We had twelve Cannon in front, four at each end, and four in the middle; . . . and behind the first line our Cohorns played; tis said the enemy intended to wait our attack, but our whole artillery played so briskly on them and galled them so terribly, that their right, some say, without order, advanced with a great fury in a highland trott in a deep column and in an unsoldier-like manner firing without order and moving sideways with their targets and broad swords as to stretch out to the length of our left wing; we kept our fire till they were near; but not withstanding, they broke the first line of Barrell's Regiment on our left, and being let in, they were flanked by them, and met by the second line in front who tis thought by their fire killed several of Barrell's mixed with the enemy.

Source: Jasper Leigh Jones, *A Plan of ye Battle of Colloden between his Majs. Forces Under the Command of his Royall Highness the Duke of Cumberland and the Sctt. Rebels April ye 16 1746* (1746).

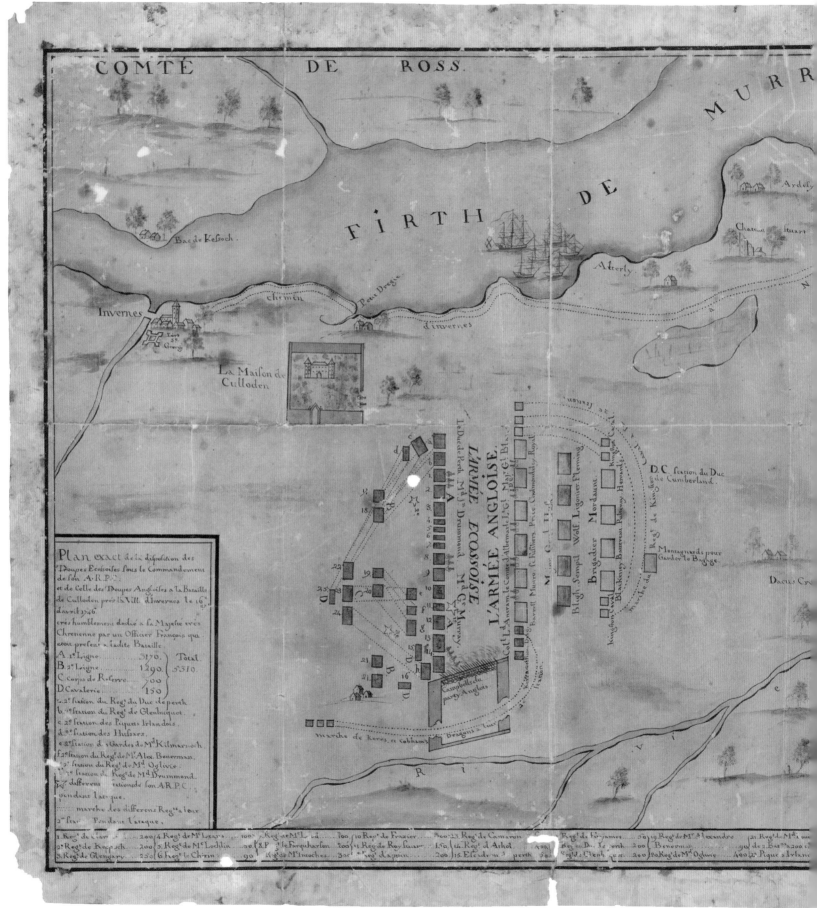

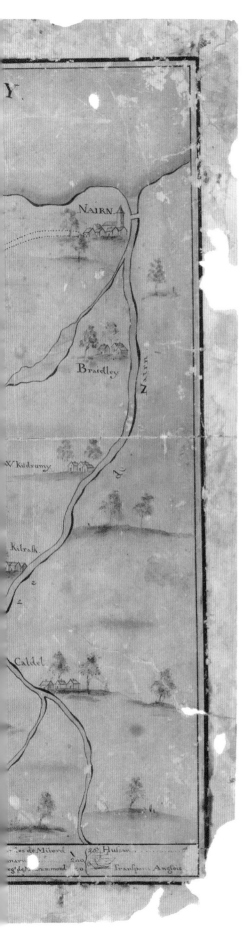

FIGURE 4.22

This manuscript map of Culloden, drawn up around 1748 by a French officer who was present on the day, gives a rare Jacobite view of the battle. The phrase 'son A.R.P.C.' in the title stands for 'son Altesse Royale Prince Charles' (his Royal Highness Prince Charles) and the map is dedicated to 'Sa Majeste tres Chretienne' (Louis XV). Charles Edward is shown no fewer than three times: in advance of the front line, behind the left flank and behind the right flank of the Jacobite army. The plan also gives a good impression of the main movements of the armies during the battle – how Cumberland wheeled his third line around to the right to counter new troops on the Jacobite left near Culloden Park, and how the Hanoverians took over the Culwhiniac enclosure to the south. It also shows how the central Jacobite advance veered to the right so that it obstructed the following regiments, squashing the Jacobite charge and bringing it to a halt. Excellent details of the armies, commanders and the various regiments and clans are also given. The map was handed down through family lines in French military circles, before being taken to the United States and then donated to the National Library of Scotland in 1996.

Source: Anon., *Plan exact de la disposition des Troupes Ecossoises sous le Commandement de son A.R.P.C. et de Celle des Troupes Angloises a la Bataille de Culloden prés la Ville d'Invernéss le 16ᵉ d'avril 1746* (1748).

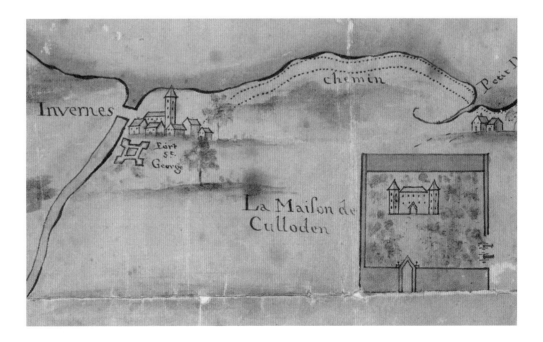

FIGURE 4.23

(*Overleaf*) John Lambertus Romer, *Fort William; Fort Augustus; Fort George, Inverness* (c.1729–46).

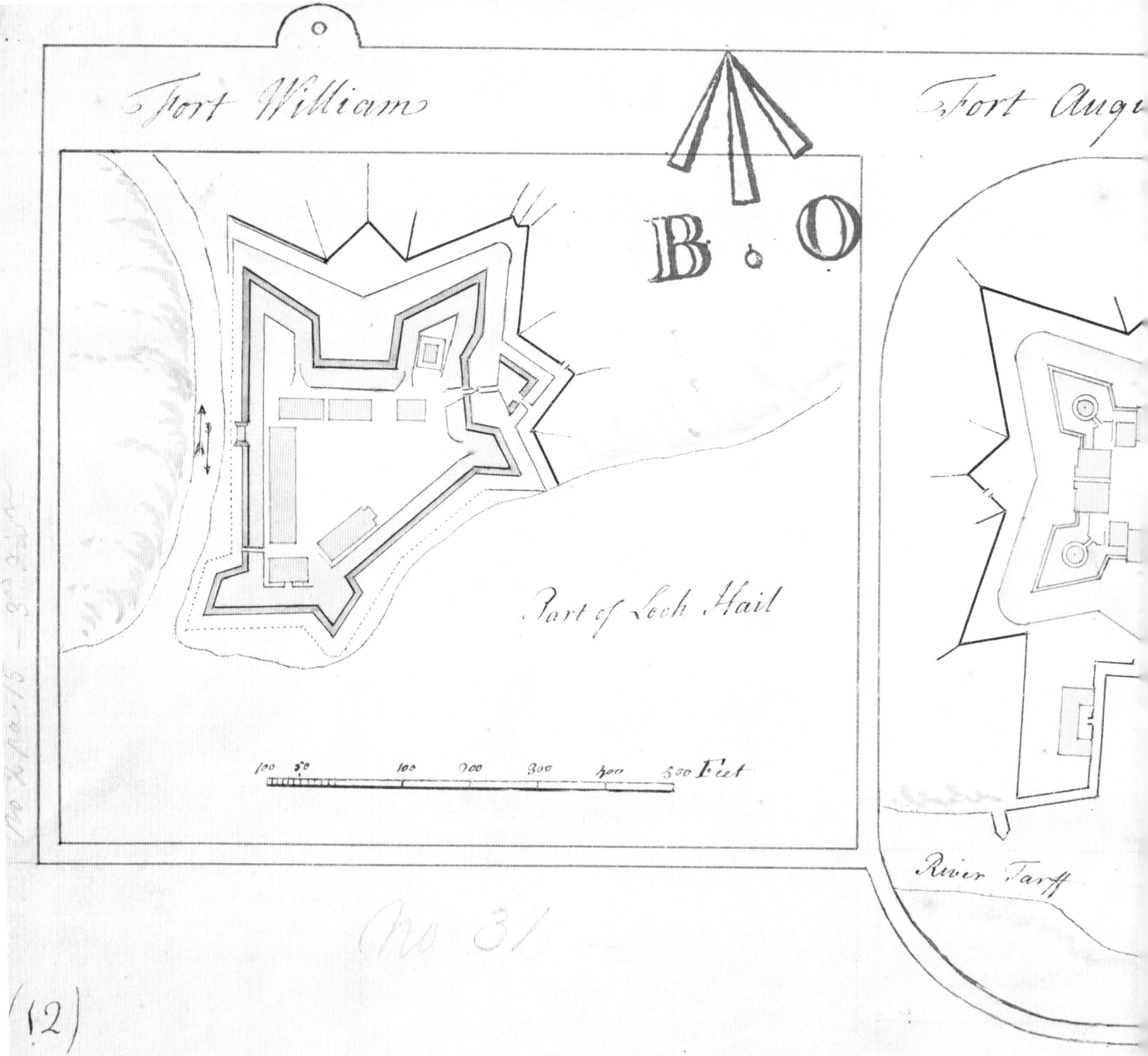

Fort William   Fort Augu[stus]

B. O.

Part of Loch Aail

100 50 100 200 300 400 500 Feet

River Tarff

(2)

Z2/31c
Fort George Inverness
Loch Ness
River Ness
Feet
5 10 30 50 70 90 200
B5 F51

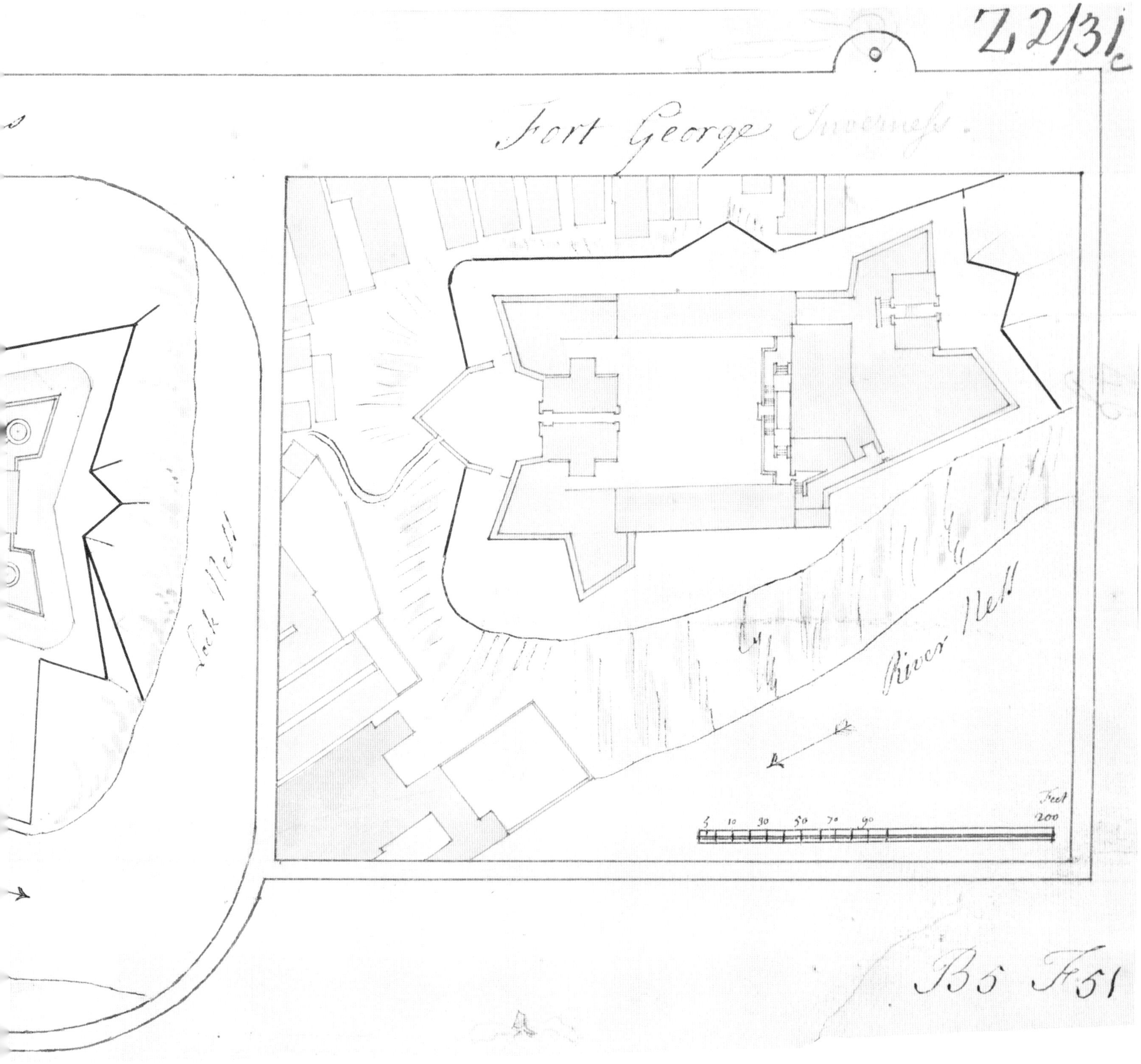

CHAPTER FIVE
THE ROY MILITARY SURVEY TO FORT GEORGE, ARDERSIER, 1746–1787

While the battle of Culloden had been a resounding victory for the Duke of Cumberland and his Hanoverian army, the 'Forty-five was an embarrassment for the British government. Two government defeats, at Prestonpans and Falkirk, and the appearance of Charles Edward Stuart and his Jacobite army within 100 miles of London, had shown tactical and strategic weaknesses in the government's military handling of Jacobitism. That only one of the Great Glen garrisons – Fort William – had withstood a Jacobite siege, while Fort Augustus and Fort George at Inverness were in ruins, was humiliating.

Jacobite hopes may have been crushed at Culloden, but there was no guarantee that the battle marked the end of Jacobitism. Cumberland was therefore determined to ensure that the Jacobites would never again pose a military threat to the security of Great Britain. Instead of returning to London to celebrate his victory, Cumberland and his army headed west, to Fort Augustus, to begin a period of brutal and indiscriminate reprisals. Anyone suspected of Jacobite sympathies was executed or imprisoned, estates were forfeited and another Disarming Act, this time banning Highlanders from wielding weapons, wearing kilts and playing bagpipes, was harshly enforced. By September 1746, there were 15,000 government troops in Scotland. Plans were once again drawn up to increase the number of garrisons in the Highlands and to expand existing barracks to provide quarters for the troops tasked with policing Scotland (fig. 5.1).

This chapter looks first at the Military Survey of Scotland overseen by William Roy as a direct consequence of events during the 1745 Rising, which had made the British government and military acutely sensible of the fact that 'the

FIGURE 5.1

(*Opposite*) After the 'Forty-five, resistance to Hanoverian rule in the Lochaber area was still strong. Anticipating further dissension, Fort William was hastily repaired and a temporary timber barracks was built on the vacant ground near the governor's house to accommodate an additional 160 men posted to police the wider Lochaber area. Before Dugal Campbell joined the Duke of Cumberland at Culloden, he had been responsible for the repair and modification of Fort William. Plans from 1744 and 1745 show that the main gate, ramparts and parapets had been rebuilt, and new officers' barracks were being built. Campbell's detailed plan and elevations of the timber barracks show the basic nature of the quarters for the soldiers. The interior features include fireplaces, partitions and eight bunks per room. Each of the ten rooms was approximately 16 by 30 feet, with two external staircases for access.

Source: Dugal Campbell, *Plan, Elevation and Sections of additional Barracks of Timber . . . to be built at Fortwilliam . . .* (1746).

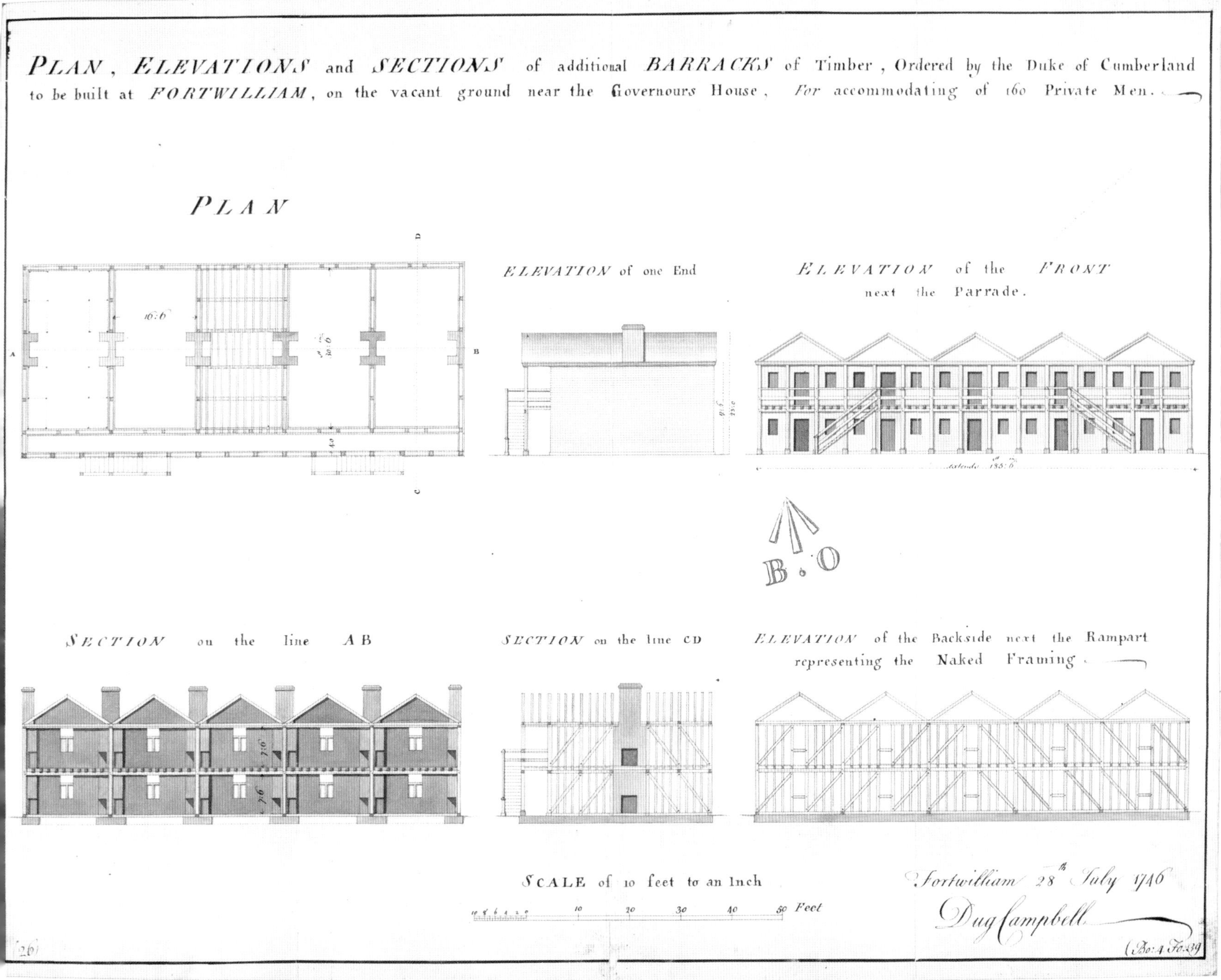

PLAN, ELEVATIONS and SECTIONS of additional BARRACKS of Timber, Ordered by the Duke of Cumberland to be built at FORTWILLIAM, on the vacant ground near the Governours House, For accommodating of 160 Private Men.

PLAN

ELEVATION of one End

ELEVATION of the FRONT next the Parrade.

SECTION on the line AB

SECTION on the line CD

ELEVATION of the Backside next the Rampart representing the Naked Framing.

SCALE of 10 feet to an Inch

Fortwilliam 28th July 1746

Dug Campbell

Geography of our Country [Scotland] is very little known'. We then look at the measures taken by the Board of Ordnance engineers to repair and expand the military garrisons in Scotland, before turning to the road-building programme, which once again became a priority in 1749 to connect the forts with each other. A fourth theme looks at the plans recording the construction of the mighty Fort George at Ardersier Point which, when completed, was the largest barracks in Britain. We conclude by looking at the new military threats to coasts and harbours from the late 1770s.

The Military Survey of Scotland, 1747–55

John Campbell, fourth Earl of Loudoun, was one of many officers tasked with punitive operations in the Highlands following the battle of Culloden. Charged with hunting down rebels in Inverness-shire, Campbell claimed his only topographical reference was to relevant pages in a well-worn copy of Blaeu's 1654 *Theatrum Orbis Terrarum sive Atlas Novus*. Most of the maps in the *Atlas Novus* were drafted around 1600; by the time the Earl of Loudoun was referring to the map of Moray and Nairn to track down Jacobites, the map was over 150 years old! Loudoun simply had rivers, lakes and settlements, woodland and stylised topography to guide him (see fig. 2.7b). Loudoun's experience was similar to Cumberland's who, with 'the Generals under his command, found themselves greatly embarassed for the want of a proper Survey of the Country'. While Cumberland and his staff set out their military strategy for enforcing the rule of law as they camped at Fort Augustus in the summer of 1746, it was the army's Deputy Quartermaster-General, Lieutenant-Colonel David Watson, who first conceived the idea of a military survey of Scotland. As William Roy, the principal surveyor on the Survey, later explained:

> The rise and progress of the rebellion which broke out in the Highlands of Scotland in 1745 . . . convinced Government of what infinite importance it would be to the State, that a country, so very inaccessible by nature, should be thoroughly explored and laid open, by establishing military posts in its inmost recesses, and carrying on roads of communication to its remotest part.

When Cumberland returned to London in July 1746, having resigned as Commander in Chief of the forces in Scotland, he presented Watson's proposal for a 'compleat and accurate Survey of Scotland'. The notion was approved and Watson was tasked with directing the Survey, assisted by William Roy.

Although proposed as a military survey with the backing of government, in its production it was poorly supported by the Board of Ordnance, both financially and with personnel. Watson wrote in June 1748 how 'the Surveying Scheme, . . . has given me Infinite Pain'. On the same day, William Skinner, Director of Engineers in Scotland, was informed that the intended number of engineers and soldiers to be employed on the Survey was being reduced. For the first two years, Roy worked alone on the Survey. In 1749, John Manson, recently promoted from the Drawing Room to Practitioner Engineer, assisted him. The following year, three junior engineers joined the survey team: Hugh Debbeig, previously employed in surveying the proposed road from Newcastle to Carlisle, and John Williams and William Dundas, both cadets from the Royal Military Academy. A month later, Thomas Howse left the Academy to join the Survey and, in 1752, David Dundas, at the age of fifteen, became an assistant surveyor in Scotland.

The Survey was conducted in two parts. The north of Scotland, including the Highlands but not the Islands, was surveyed between 1747 and 1752, resulting in an 'original protraction' and a 'fair copy' of the composite map (fig. 5.2). Between 1752 and 1755, the Lowlands were surveyed and a 'protracted copy' of the south of Scotland was produced (fig. 5.3); no fair copy of the Lowlands was made. Surveying took place during the summer months. In the autumn, the surveyors returned to the Drawing Room in Edinburgh Castle and, through the winter, the separate traverses were collated into a single map (the 'original protraction'). While the survey parties were responsible for recording and sketching data in the field,

Paul Sandby, acknowledged as the 'father' of English watercolour art, was the 'chief Draftsman of the fair Plan'. Two further draughtsmen completed the full complement of artists – Charles Tarrant, whose exceptional cartographic style was highly commended by the Board of Ordnance (figs 1.9, 5.5b, 5.7b), and John Pleydell.

Some of the most detailed methods of surveying in eighteenth-century Scotland come from documents relating to the Military Survey. The Survey, although carried out at a time of relative peace in Scotland, was the work of rapid reconnaissance rather than a comprehensively measured topographical survey. The surveyors worked along sets of traverses using basic theodolites (or circumferentors) to measure angles (fig. 5.4), and iron chains of 45 or 50 feet to measure distances. With these 'common' instruments, 'the courses of all rivers and numerous streams were followed and measured; also all the roads and the many lakes of salt-water and fresh'. Other points were fixed by the intersections of bearings taken from traverse stations, and the remaining landscape features – towns and settlements, enclosures and woodland, and relief – were sketched in by eye, not accurately measured.

In 1755, work on the Military Survey of Scotland came to an end when the Survey personnel were called away to the impending Seven Years' War between Britain and France (1756–63). But for this intervention, the Survey would have been completed and, according to Roy, 'many of its imperfections no doubt remedied' because 'being still defective, points out the necessity of something more accurate being undertaken, when times and circumstances may favour the design'.

With the outbreak of the Seven Years' War came renewed Jacobite hopes of an invasion of England and a corresponding uprising in Scotland by Irish and Scottish 'Rebel' regiments. In a letter to the Lord Advocate, John Forbes wrote: 'I am very well Convinced That none of them will be very rash to engage in anoyr Rebellion, tho' I should be very sorry if this part of the Country (I mean the highlands) be not look'd very well after & some Regts kept Constantly among them'. If an uprising was to occur, Forbes was adamant that David Watson, owing to 'his knowledge & ability sp[ent] in the manadgement of affairs in Scotland which he has made more his study than any man alive [and] knows every Corner of the Kingdom', should be dispatched back to Scotland. There is no evidence to confirm the Survey was ever put to military use. When Watson died in 1761, the fair copy of northern Scotland and

FIGURE 5.2

(*Overleaf*) David Watson wrote instructions to guide junior engineers in their surveying work. These may have been written for the engineers working on the Military Survey or as a result of the Survey – the instructions are not dated. In his 'Orders and Instructions to be Observed by Assistants, in Reconnaitring, Examining, Describing, Representing and Reporting, any Country, District, or particular Spot of Ground', Watson focused on military functionality from the outset. He explained that:

> As the Encampments, Marches, and every possible movement proper for an Army to make in the Field, entirely depend on a just and thorough knowledge of the Country, the greatest care & Exactness should be observed in Examining minutely the Face of that Country.

It was important for a surveyor to record the land use and the nature of the terrain, whether it was impassable or could be traversed by foot or by horse; to be exact in describing the location and size of settlements; to mark all rivers and lakes, as 'the nature of any River or water . . . are allways of the greatest Consequence to Troops in Time of Service'; and to show hills or high ground. Watson explicitly instructed that the state of the roads, their widths and distances between destinations, should be recorded.

When comparing these instructions to the Survey, it can be seen that some of the specifics may not have been applied. Roads, for example, are depicted with variable accuracy and inconsistently from one sheet to the next. Proposed roads or roads under construction would naturally pose challenges as to what to record, but some roads known to exist stop at the edge of a sheet and fail to continue onto the next – the road south-west of Culloden, for example. While some features that were known to exist were excluded, others such as the mighty Fort George at Ardersier are shown as complete, when construction had not even begun. The drawing specification also varies. As this extract from the Survey of the area around Dunkeld shows, the military road running north–south (from Perth to Fort William and Inverness) is shown as two parallel lines, while the road west, to Amulree, is shown as a single line.

Source: William Roy, Detail from Dunkeld environs, Roy Military Survey of Scotland (1747–52). Courtesy of the British Library Board.

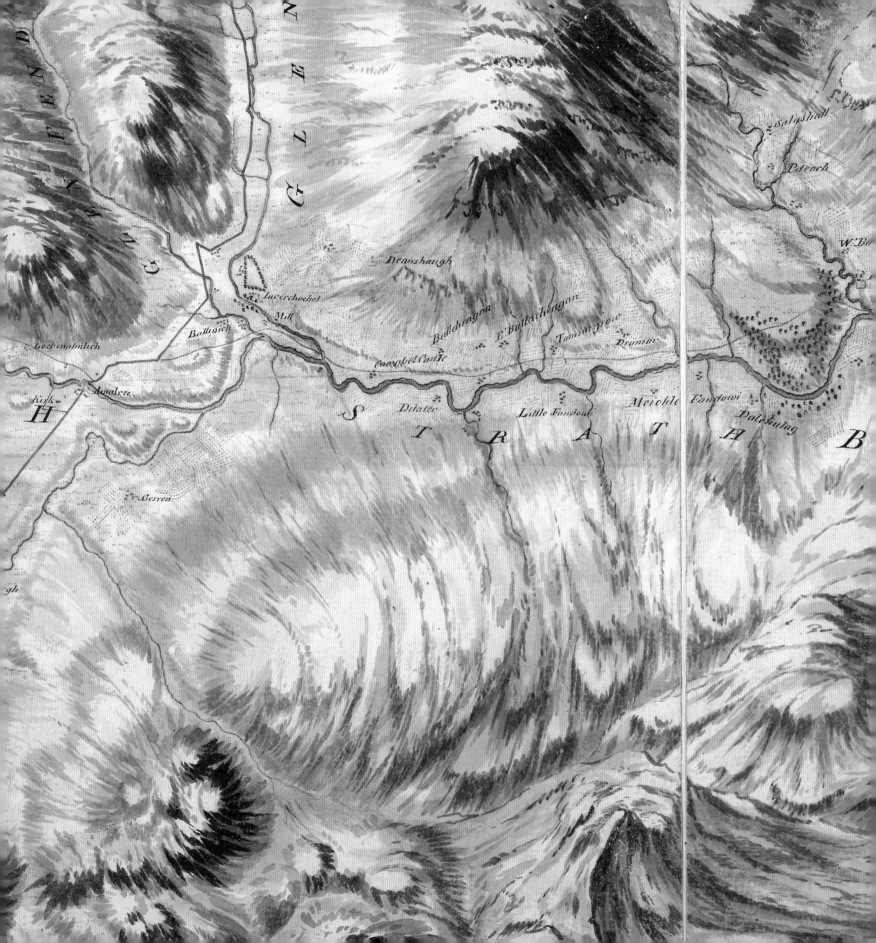

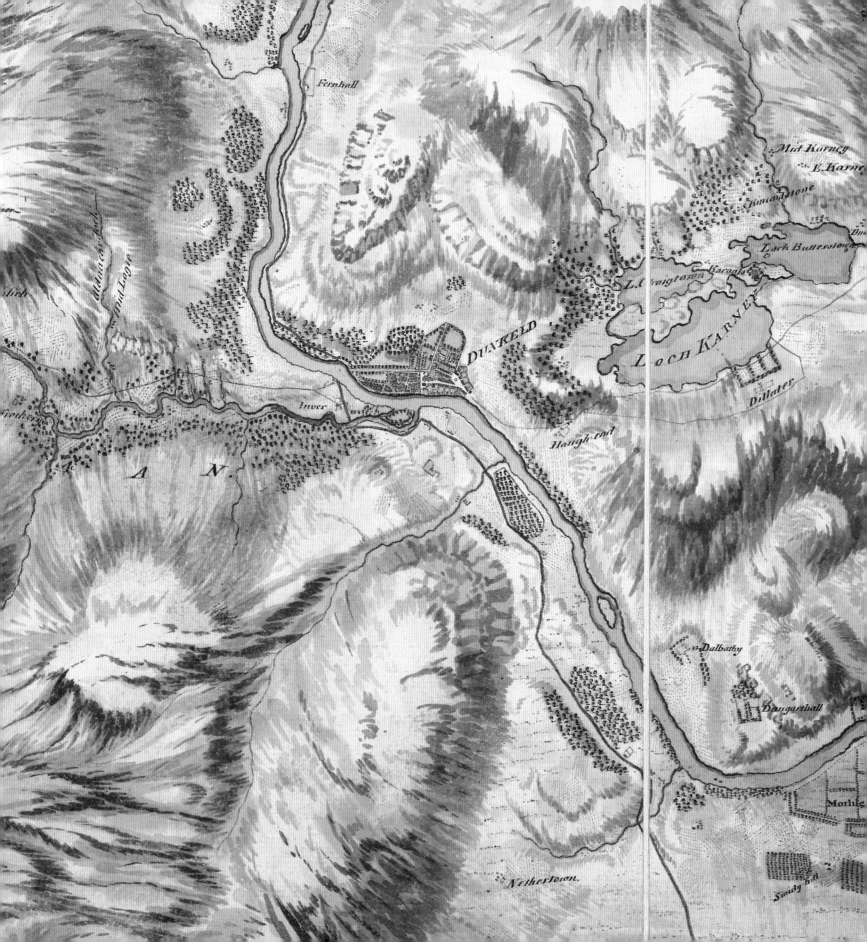

FIGURE 5.3

(*Opposite*) The difference between the fair copy (a) and the original protraction (b) is striking. As this image of the area around Stirling shows, the original protraction was drawn for the most part in black ink with very few features in colour. The fair copy, in contrast, abounds with colours that reflect the prevailing military colour schemes and conventions that had emerged during the eighteenth century: red was used for buildings and man-made structures, brown for roads, blue-green for water, green for woodland, yellow for cultivated ground, and buff for moorland. Hill features were drawn in the emerging style of the time, using hachured lines to indicate the direction of the slope and changing tones to differentiate the gradient (see fig. 5.2). The final 'relief' compilation resulted in a hybrid combination of bird's-eye and perspective views. The relatively few symbols are stylised representations of trees, tilled fields, moorland, and sands or shoals. Other features such as mills and churches were shown sparingly. Notably, the maps were not graduated for latitude and longitude, nor presented with a scale statement. Aaron Arrowsmith determined their scale as one inch to 1,000 yards (1:36,000).

The overall effect was to produce a highly painterly and decorative interpretation of contemporary military cartographic techniques. William Roy later described the unfinished manuscript as rather a 'magnificent military sketch than a very accurate map of a country' but, although an 'imperfect work', it still possessed 'considerable merit, and perfectly answered the purpose for which it was originally intended'. Hugh Debbeig, one of the Survey's contributing engineers, who clearly felt no need for modesty, referred to it as 'the greatest work of this sort ever performed by British Subjects and perhaps for the fine Representations of the Country not to equal in the World'. The map is a snapshot of mid eighteenth-century Scotland and was intended as a political tool. For many parts of the Highlands, and northern Scotland in particular, it is the only relatively large-scale topographical map in existence for the eighteenth century.

Source: William Roy, Details from Stirling environs, Military Survey of Scotland (1747–55).

FIGURE 5.4

This is an example of the type of basic theodolite or circumferentor that was probably used to measure bearings on the Roy Military Survey. The circumferentors were fairly simple: a graduated disc around 7 inches in diameter with a magnetic 'needle box' and an alidade for measuring angles.

Source: John Adams, 'Image of a circumferentor, similar to those used on the Roy Military Survey', in *The Elements of Useful Knowledge . . .*, 2nd edition, Plate XV (1799).

the protracted copy of southern Scotland were found among his possessions. No doubt Watson would have made good use of these copies if he had been recalled to Scotland.

Refortification

Much was needed to be done to repair and expand the fortifications in Scotland following the 'Forty-five. Fort William was hastily renovated. Fort Augustus was restored and reported as being in 'good order' by May 1750. Fort George at Inverness, however, was 'intirely in Ruine' and beyond repair (fig. 1.9), so an alternative solution had to be found. One option was to abandon Castle Hill and rebuild Cromwell's pentagonal citadel down by the harbour. In 1746, Major Lewis Marcell, an Irish engineer, surveyed and compiled a set of plans for a new fort to be 'Done exactly upon the Old Lines of Olivers Fort'. The following year, William Skinner, the new Chief Engineer for the Board of Ordnance in Scotland, was ordered to review the site, to consider its situation and to 'make proper plans, sections, and Elevations of the adjacent Ground within the reach of Cannon shot' (fig. 5.5). Skinner estimated the cost of building a new fort on this site at nearly £93,000, the high cost in part due to limited building resources within easy access of 'Oliver's Fort'. Inverness Town Council, who owned the ground on which the fort sat, was also demanding a fee for the land and compensation for the loss of the harbour (£3,000). Ultimately, the decision was made to look elsewhere for a suitable site.

As well as carrying out repairs to existing forts, the military engineers were also to reconnoitre the country and to look out for places to establish military garrisons. William Keppel, 2nd Earl of Albemarle, who commanded forces in Scotland after Cumberland, was at a loss to know how all this could be achieved. As he wrote in February 1747 to the Secretary of State, Thomas Pelham-Holles:

> The scheme your Grace is pleased to send me, for erecting other Forts or stations in different Parts of the

FIGURE 5.5

(a) (*Opposite*) Lewis Marcell completed a very thorough survey of the original Cromwellian citadel, 'Oliver's Fort' at Inverness, and supplied multiple plans to show the governor's house (shown here), barracks for two battalions of 700 men each, a powder magazine, bakehouse, brewery, stables, hospital, armoury and arsenal. Aware that the final cost for building the fort might be considered too high, he noted that 'The Buildings in the Angles Shaded Yellow . . . are thought Requisite for Obtaining all the Conveniency's necessary to the Fort. But are distinguished by this colour . . . so they may be rejected if judged Superfluous.'

(b) (*Below*) William Skinner's plan, drawn by Charles Tarrant, is very similar to Marcell's 'General Plan' of the fort (1746), both of which follow the pentagonal lines of the original citadel, encircled by a moat of water drawn from the River Ness. Skinner's report offers one of the few references to methods of survey in the Board of Ordnance records. He advised that 'As soon as the Weather permitts, [I shall] begin a Survey of the old Remains . . . in order to make my designe's and shall Employ a few men to pen the angles that I may fix its present Situation, and try if any of the Remains of its former Foundations are to be traced.'

Sources: (a) Lewis Marcell, *The Governours, Deputy Governour's, Fort Major's & Storekeeper's Houses; Marked Nos. 1 & 2 on the General Plan* (1746). (b) William Skinner and Charles Tarrant, *Plan for Building a Fort at Inverness on the Vestige of an Old Fort Demolished 1746* (1747).

(b)

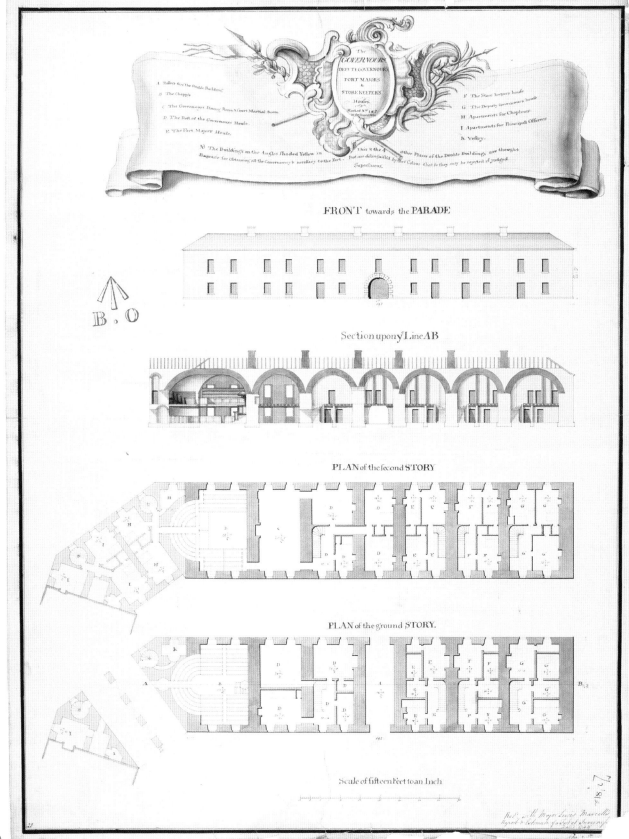
(a)

SCOTLAND: DEFENDING THE NATION

Country may be put to very good use... in preventing the Incursions or Depredations of the Highlanders, But how to carry on the building of Five, or Six of those Forts sufficient to contain Ten, or Twelve Companies Each at the same time that the Fort at Inverness is erecting and those of Fort Augustus & Fort William repairing with Numbers of Hands & at a vast Expence, I am at a loss how to execute; And in my Opinion it is rather the Work of Years, than of one Summer.

Albemarle was placed in temporary command of the engineers in Scotland and, during this brief period (1746–47), reconstructions of fortifications and commissions for maps and plans proceeded unhindered by financial constraints. David Watson assisted Skinner in surveying suitable locations and structures for establishing military detachments in the Highlands, accompanied by Paul Sandby and Charles Tarrant, who between them compiled some stunning pen, wash and watercolour plans and perspective views of Scotland's castles (fig. 5.6).

By this time, the medieval castles of Corgarff and Braemar, surveyed by Watson and ordered by the Duke of Cumberland to be put into 'a Condition fitt to accommodate His Majesty's Forces', were converted into barracks (fig. 5.7). The Board of Ordnance compulsorily took over these two castles, as well as Tarbat and Tioram, not for crimes committed by their owners but following a report by the Duke of Montagu, Master-General of the Ordnance, stating that it was 'necessary for the Tranquillity of that part of the Kingdom that the said Four Castles should be [taken] Possession of [and to] cause them to be repaired, refitted, converted and supplyed'. Repairs were also made to the four barrack forts – Inversnaid, Kiliwhimen, Bernera and Ruthven – 'at very little cost'.

The garrisons in the Lowlands – Edinburgh, Stirling, Dumbarton and Blackness – were also increased following the 'Forty-five. At Edinburgh Castle, the Long Storehouse was converted into barracks and a new Barrack Hall for a further 270 men was designed by William Skinner in 1750. In 1755, a new barracks was built along the north side of Parade Yard. A new powder magazine was built at an estimated cost of

FIGURE 5.6

(*Opposite*) Paul Sandby's masterful depiction of Dumbarton Castle is one of the most artistically accomplished of all the Board of Ordnance's graphic representations of Scotland's forts. From the time of Leonardo da Vinci (1452–1519), both planimetric (overhead) and ground-level views of the same place had proved their value, particularly for military engineering. Sandby, however, goes a stage further in setting his planimetric view inside a subtle bird's-eye front and rear view of the castle. Dumbarton's rugged and craggy mass of rock, with its prominent central gully, is brought to life, with the views from the west and east forming an aesthetically pleasing and informative whole.

Dumbarton has a long history as a fortified site. Recent refortification following the 1715 Rising included the construction of the governor's house and King George's Battery, guarding the south-east entrance (shown at the bottom of the map and in the lower-left view). The old entrance and Wallace Tower (north-west side), dating from the early seventeenth century, are shown in the lower-right view. Further work was planned in 1748, including the construction of a new gunpowder magazine; new gun batteries and a French prisoner-of-war building were constructed in the 1790s. Sandby's map was held by William Skinner and eventually descended to Monier Skinner, who passed it to the War Office in 1872. By this time, the familiar red B.O. (Board of Ordnance) arrowhead stamp of the eighteenth century had changed to the Inspector-General of Fortifications (I.G.F.) stamp, seen here.

Source: Paul Sandby, *Plan of the Castle of Dunbarton* (c.1747).

REFERENCES

PLAN
of the CASTLE of
Dunbarton

Z 3/57

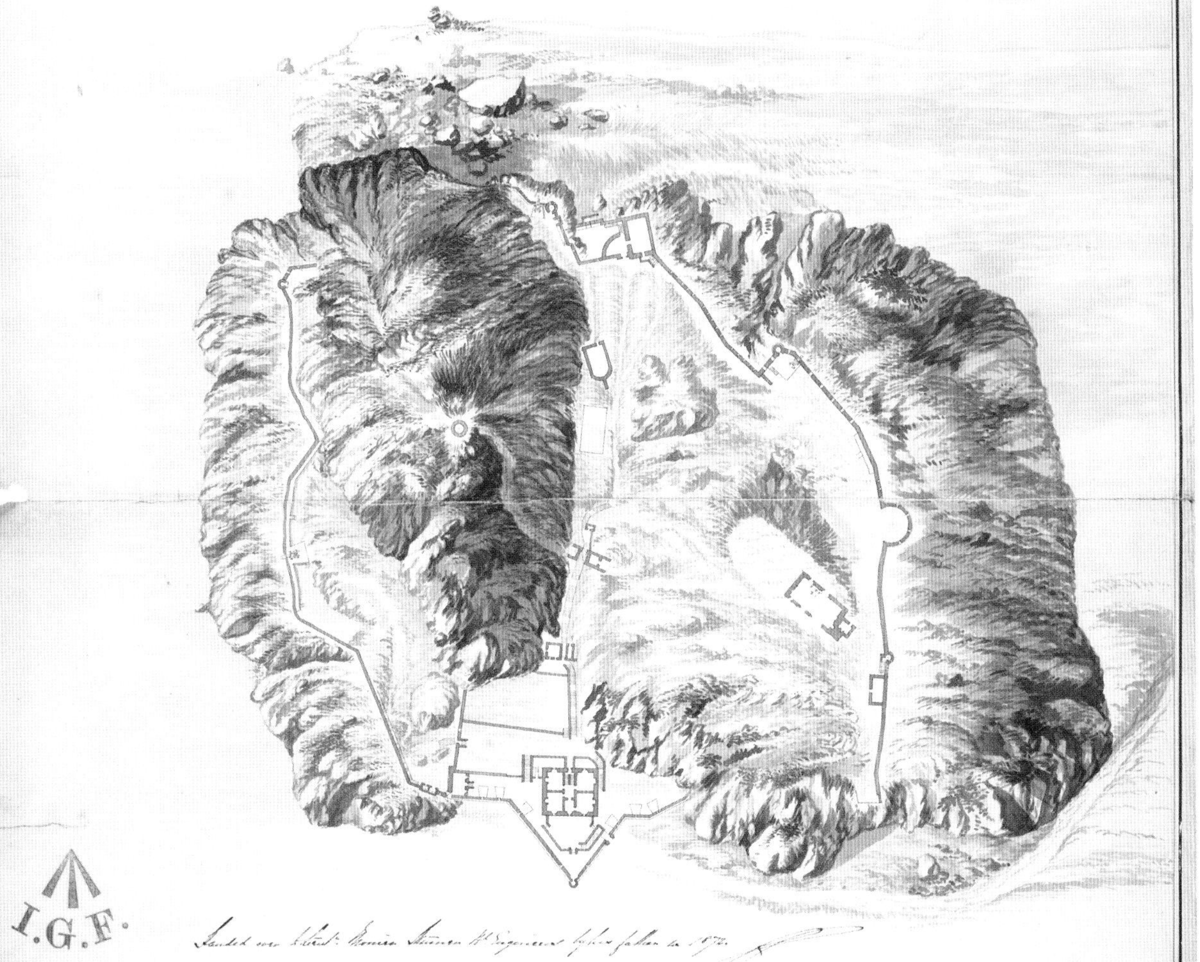

I.G.F.

(a)

Old PLAN of CORGARFF CASTLE.

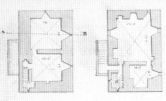    

Repairs

   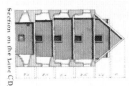

Scale of Feet

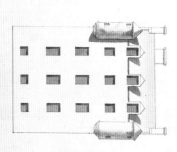 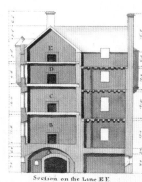 

REPAIRS of BRAEMARR CASTLE

 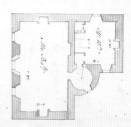 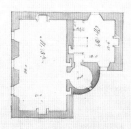 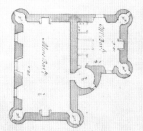

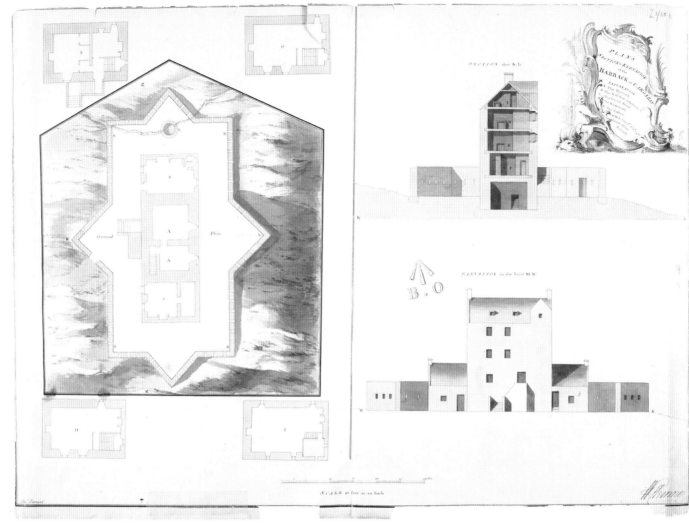

FIGURE 5.7

Corgarff Castle lies approximately fifteen miles north-west of Ballater, in the upper reaches of Strathdon. It is an attractive but lonely and windswept spot, its remoteness making it an ideal base for fomenting disaffection – in 1715, the Earl of Mar had marched from his ancestral seat at Kildrummy, further down Strathdon, via Corgarff to Braemar, recruiting and arming his Jacobite force *en route*. Braemar Castle, on the south bank of the River Dee, is approximately twelve miles to the south-west of Corgarff. Built in 1628 for the Earl of Mar, this battlemented tower house was burnt down by Jacobite forces in the 1689 Rising, the 5th Earl of Mar having joined the new government under William and Mary. The castle was rebuilt, only to then fall into ruin following the 1715 Rising. Both castles lie on the military road linking Blairgowrie with Fort George at Ardersier, which was constructed between 1749 and 1753.

(a) (*Opposite*) In 1748, the medieval tower houses were requisitioned by the Board of Ordnance and converted into soldiers' barracks – Corgarff to accommodate an officer and forty men, and Braemar an officer and fifty men. Corgarff was gutted and another floor inserted; the three upper floors, shown in the vertical section here, became barrack rooms, each intended to hold eight double beds. The upper floors of Braemar were repaired, with each designed to hold fourteen beds. The officers, one to each barrack room, had a bed to themselves, while the privates slept two to a bed, although half the garrison was expected to be out on patrol at any one time. While the ground floor of Braemar Castle was large enough to provide a storeroom, kitchen and prison, Corgarff was extended on either side of the tower to provide a guardroom, prison, bakehouse and brewhouse with storerooms in between.

(b) (*Above*) At both towers, an eight-pointed curtain wall was constructed around the perimeter, with gun-loops to take muskets. The walls were intended to provide protection against a lightly armed attack only, not one with artillery.

Source: (a) Anon., *Old Plan of Corgarff Castle; Repairs; Repairs of Braemarr Castle* (1749). (b) William Skinner and Charles Tarrant, *Plans Section & Elevation of the Barrack at Cargarff* (c.1750).

£1,495 in readiness for the increased military strength of the garrison (fig. 5.8). The magazine, built on the site of the old one, was commissioned after a complaint to the Board of Ordnance that 'the Magazine at Edinburgh was insufficient to contain a proper Quantity of Powder for the Service of that Garrison'.

Caulfeild's roads, 1749–67

If the Highlands were to be 'thoroughly explored and laid open', then more roads were needed to facilitate the movement of men and machinery through the glens and over the mountains. From 1749, there was a substantial deployment of manpower in the Highlands charged with 'carrying on the Roads in North Britain', the government having finally realised that controlling Scotland required both money and men. Under the direction of Major William Caulfeild, Inspector of the New Roads and Bridges, four engineers – Harry Gordon, James Bramham, George Morrison and George Campbell – oversaw the building and repair of military roads and bridges, and mapped their progress. The workforce was formed of approximately 1,350 men from five regiments.

Two main roads through the Highlands were constructed. The first ran in a north-westerly direction from Stirling to Fort William. Construction began in 1749, with men from Colonel Pulteney's Regiment working on the section between Loch Lubnaig, near Callander, to the start of Lochan Lairig Cheile in Glen Ogle (fig. 5.9). At the same time, a detachment from Lord Sackville's Regiment laid gravel and mended the section from Callander to Loch Lubnaig. Maps of the continuation of the road northwards, from Loch Dochart at the foot of Ben More, past Crianlarich and Tyndrum, to Bridge of Orchy and Loch Tulla, were drawn in 1750 and 1751. Working southward from Fort William, detachments from Colonel Rich's and General Guise's Regiments constructed the road as far as Kinlochleven in 1750 (fig. 5.10).

The second main road ran north–south, between Blairgowrie and Braemar. In 1749, a detachment of Lord Bury's Regiment joined 300 men from the Royal Welsh Fusiliers to work on the road northwards from Blairgowrie. Concurrently, a detachment from General Guise's Regiment worked southwards from Braemar. The intention was that the parties would meet at the Spittal of Glenshee, but both finished the season short of the destination. In 1750, the remaining six miles were completed (fig. 5.11). Later that year, Caulfeild proposed extending the road northwards, first to Corgarff, then northwest via Grantown-on-Spey and Dulsie to Fort George on Ardersier Point. George Campbell, Engineer and Cadet Gunner, completed several surveys to accompany Caulfeild's report. The road is shown on the Military Survey of Scotland, which indicates that it was under construction in 1752; it was completed in 1754. A road linking Bernera to Fort Augustus was begun in 1755, and one from Inveraray to Tyndrum in 1758. In the 1760s, a military road was built westwards, from Bridge of Sark to Portpatrick (Dumfries and Galloway) to assist in the movement of troops to Ireland. Under Caulfeild's direction, approximately 608 miles of roads were built and a further 223 miles were under construction in Scotland by the time of his death in 1767. In his penultimate year, he directed the repair of 270 miles of road and the rebuilding of four bridges.

From the outset, the 'King's Roads' were surveyed and built for strategic reasons. Politically, the roads were a means of access, and the maps a means of imagining and visualising routeways from afar in order to impose military order. Militarily, the road network meant infantry, artillery, munitions and stores could be moved through the Highlands. The continued strategic purpose of Wade and Caulfeild's military roads was highlighted in 1808 when contingency plans were being drawn up to counter a Napoleonic invasion. The military concern was to redeploy troops from Inverness and Fort George at Ardersier to the south of Scotland and to consider strategies for concentrating the forces in the north to protect the Moray Firth. A report 'relative to the Routes by which Troops may march' identified three routes – 'distinct and separate from each other' – by which to travel. The first was the main military road through the Highlands, from Inverness and Fort George by way of Aviemore and Dalnacardoch and then to Perth, Stirling or Glasgow. The planned route was by way of Wade's military roads, at least as far as Perth and Stirling (see fig. 4.11). The second route was from Inverness and Fort George to Grantown, Braemar and on to Perth: one of Caulfeild's military roads. The third route was by way of Nairn, Fochabers and Huntly to Aberdeen; in parts, it was Caulfeild's military road, too. The report concludes: 'With regard to the resources and improvement of Scotland, it may be satisfactory to state, that troops may now march conveniently through many parts of the Country.' The roads had been considerably improved since Wade and Caulfeild's time but, this aside, the strategic vision and meticulous planning behind routes to open up the Highlands cannot be denied.

Fort George at Ardersier, 1747–70

Having rejected proposals to reconstruct 'Oliver's Fort' at Inverness, and knowing that Castle Hill was beyond repair, William Skinner proposed a new site and a new design for a fort at the top of the Great Glen in 1747. This was to become Fort George at Ardersier Point (fig. 5.12). In the process of its construction, a great many plans and sections were drawn of the fort and its location (fig. 5.13). William Skinner's designs for Fort George epitomised the art of fortification that had evolved in Britain during the eighteenth century. Fort George was a model of geometric bastion architecture, allowing convergent lines of fire from the ramparts, with huge and thick low walls, and extensive outer defensive works. This massive fort covers 42 hectares of land and was built to house 1,600 men (two infantry divisions).

The pattern of construction provides an indication of the priorities of defence. The greatest perceived threat was always from the land, from an uprising in Scotland. The first line of defence to be erected was the covered way (a defensive passageway) and glacis (a bank sloping down from the fort), protecting the landward approach to the fort. In 1754, attention turned to the Point Battery, revealing a shift in defence priorities with the strengthening of the seaward batteries. Construction on the two enormous barrack buildings began in 1753; they were completed in 1764. The grand gunpowder magazine was built between 1757 and 1759 and followed a well-established design, with a thick, brick-vaulted roof strong enough to withstand a direct hit from artillery. Ordnance and provision stores, and housing for the garrison baker and brewer followed. The final building to be built was the chapel, positioned near the Point Battery. This was the only building that Skinner added to his original design. Fort George at Ardersier Point took twenty-three years to build and would have taken longer if the Board had not cut off funds in September 1770.

By the time Fort George was built, its original purpose – to provide a secure base against the Jacobite threat – had ceased to exist. Thereafter, the fort became a training facility for soldiers recruited to fight in the French and American wars until fears of French re-armament in the nineteenth century established Fort George as one of Britain's crucial coastal batteries. Until this time, only minor repairs and adjustments to the fort were undertaken (fig. 5.14).

American Revolutionary War and coastal defences

With the decline of the Jacobite threat in the second half of the eighteenth century, the garrisons in the Highlands and Lowlands were reduced, but new threats from the sea made further demands on the Board of Ordnance. On 17 September 1779, the Scots-born American privateer John Paul Jones and his squadron caused great panic when they entered the Firth of Forth. The townspeople of Kirkcaldy were put in a state of high alarm, and when the ships sailed on south to be visible from Edinburgh and Leith, drums rolled and bugles piped summoning men to hurriedly arm themselves – though generally with nothing better than pikes and claymores. The capital seemed defenceless and unprepared. Fortunately for Edinburgh, however, a gale whipped up from the west and Jones was forced to retreat, only lingering a little off the coast (a)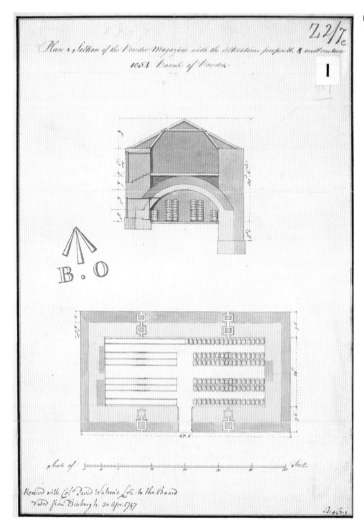

(b)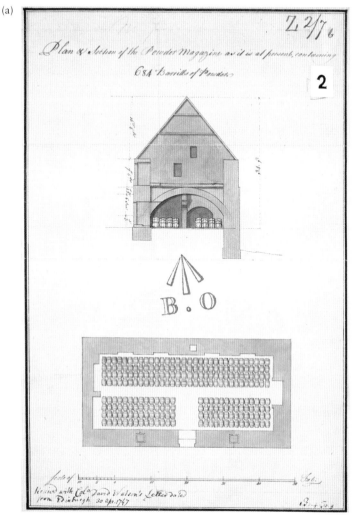

FIGURE 5.8

(a), (b) These two plans and profiles by David Watson show the proposed upgrades in April 1747 to Edinburgh Castle's gunpowder magazine, which lay to the far west of the castle rock. The construction of a new magazine increased its capacity from 684 to 1,054 barrels by extending the floor area and the span and spring of the arch. Watson also proposed thickening the exterior walls, from 1.2 metres to 2.28 metres. Watson's plan shows the distinction between the existing structure, coloured red, and the new proposed walls, coloured yellow. Good ventilation to keep the powder dry and plinths to raise the barrels off the floor are also shown. The powder magazine was built the following year, and additional storehouses were constructed to join it in 1753–54.

(c) On this 1754 plan by William Skinner, orientated to the west, the powder magazine is shown at the top (A), with the later storehouses (B and C) flanking it and facing each other. Skinner's plan also shows the location of the 'blast' wall, just to the east of a magazine, which was raised at this time. The powder magazine survived until 1890, when it was demolished at the time the storehouses became a military hospital; these buildings now house the National War Museum.

Sources: (a) David Watson, *Plan & Section of the Powder Magazine as it is at Present, Containing 684 Barrills of Powder* (1747). (b) David Watson, *Plan & Section of the Powder Magazine with the Alterations Propos'd* (1747). (c) William Skinner, *General Plan, Sections and Elevations of the Powder Magazine and Storehouses, Built in Edinburgh Castle Ano: 1753 & 1754* (1754).

(c)

GENERAL
PLAN SECTIONS and ELEVATIONS
of the
POWDER MAGAZINE *and* STOREHOUSES
Built in
EDINBURGH CASTLE
Anº 1753 & 1754

A. *Powder Magazine*
B. *Storehouse built 1753*
C. *Storehouse built 1754*

Section a.a.

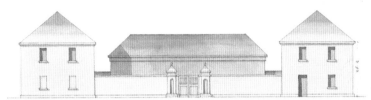

Elevation

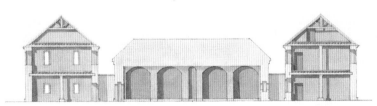

Section b.b.

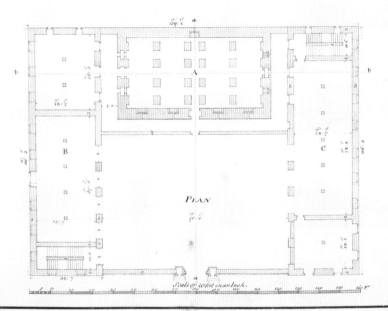

Plan

Scale of 20 feet to an Inch.

SURVEY of Part of the ROAD from STERLING to Fort WILLIAM
Made by the Party of Genl. PULTNEYS Regiment in 1749 by G. Morrison Engineer

A. Bridge of Langilay a 25 Foot Arch.
B. Bridge of Cowl 25 Foot Arch.
C. Three Bridges of 8 Foot Arch's each.
D. Bridge of Cearn 20 Foot Arch.
E. Bridge of Inverioch 20 Foot Arch.

From a to b the Way the Road is design'd. c Old Roads.
The Red Line along the Road, is where there is a Stone Wall of two & three Foot height.
The Black Lines crossing the Road, are Drains in Boggy Ground.

Laid before the Board 5 January 1749

Scale of 400 Yards to an Inch.

FIGURE 5.9

William Caulfeild expected his engineers to be 'thoroughly acquainted with the Country and its several Passes & Rivers' and to provide a description of the landscape to accompany 'an exact Plan of the Road carried on under their Inspection'. The plan was to be made at the end of every road-building season (May to September). In his report to accompany this plan, George Morrison explains that 'allmost a Mile from Balquidder [Balquhidder], is the pass of Seuii over the Hill, which is so High as to be covered with Snow [for a] great part of the Year, and the discent on the North side very difficult: these dificultys prevented the Roads being carried that way which would have been nearer' (see detail from main map). He also describes the ground conditions which are 'mostly either very Rough and Stoney, or Boggy . . . the Hills on each side are in most places very steep'. The depth of the Water of Strathyre varies from 8 to 20 feet, while all the other rivers are no more than 6 feet deep, but the river banks are mostly high and rocky. Harry Gordon, who surveyed the road in 1751, explained that several bridges were built over the burns coming down from the hills, which in summer had hardly any water but in winter were raging torrents. Retaining walls were also built, although 'tedious and Expensive', to prevent earth and stone from the banks falling onto the road, and along precipices to prevent any carriages that overturned from rolling down the steep slopes.

Source: George Morrison, *Survey of Part of the Road from Sterling to Fort William* . . . (1749).

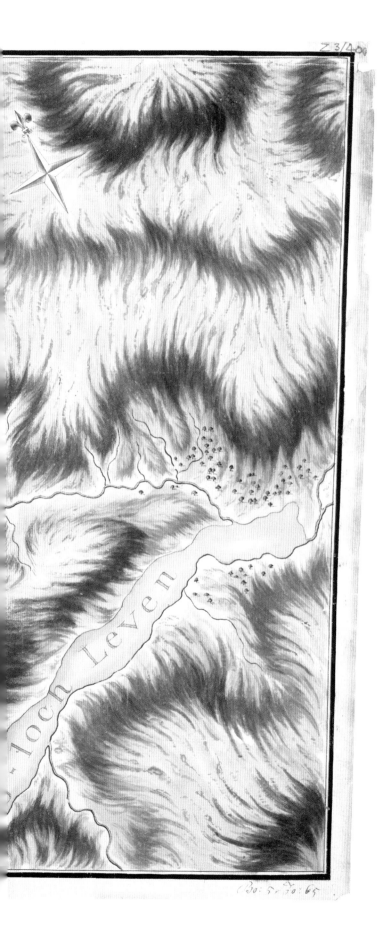

FIGURE 5.10

This map records the state of the military road under construction between Kinlochleven and Fort William in 1750. Through his striking use of hachured lines or 'centipedes' to show relief, John Archer's map shows the dramatic, and often difficult, landscape through which the military regiments built new roads. As the detail from the main map shows, Archer clearly distinguishes the sections of road made by Colonel Rich's Regiment (in red) and the section made by General Guise's Regiment (in yellow). Rich's men worked from the north, building the road through Blàr a' Chaorainn along the north bank of the River Kiachnish. Guise's men took over to construct a section of the road passing the north end of Lochan Lùnn Dà-Bhrà, before Rich's men continued the build through Lairigmòr towards Kinlochleven. Several bridges were built along the route, ranging from 10-foot to 24-foot spans. Without a bridge, crossing rivers could be treacherous in the Highlands, and waiting for water levels to drop could severely hamper the quick and easy passage of troops.

Source: John Archer, *A Survey of the Road made by Coll Rich's & Genl Guise's Regts between Fort William & the Head of King-loch Leven* (1750).

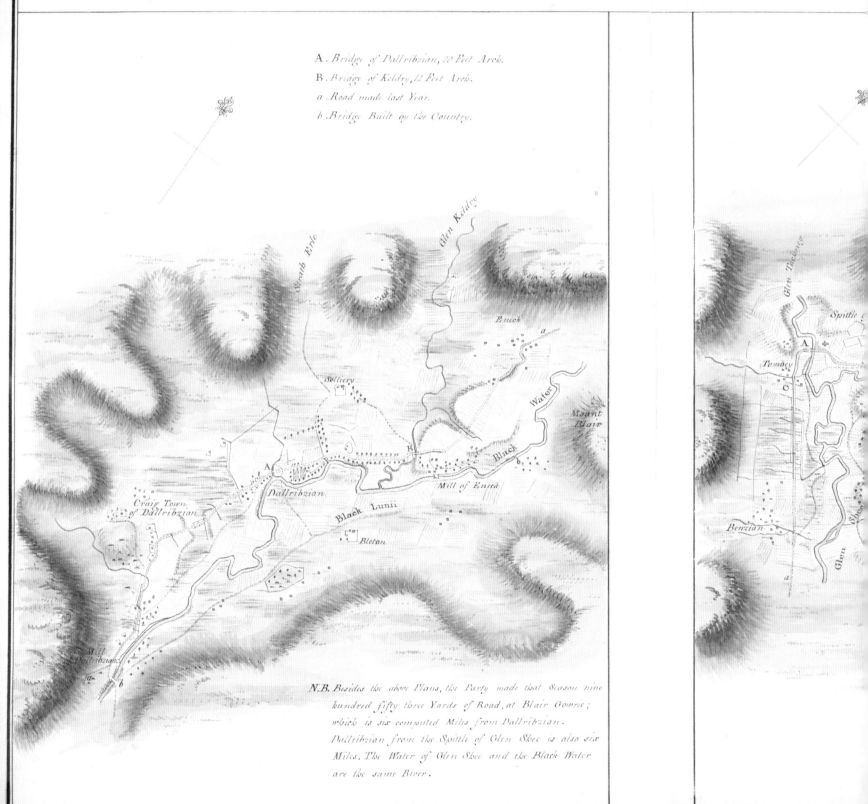

SURVEY of the different Parts of the ROAD joyned between H[...] Made by a Detachment of one hundred Men of *Lord Vis.t BURY*'s R[...]

A. Bridge of Dallribzian, 20 Feet Arch.
B. Bridge of Keldry, 12 Feet Arch.
a. Road made last Year.
b. Bridge Built by the Country.

N.B. Besides the above Plans, the Party made that Season nine hundred fifty three Yards of Road, at Blair Gowrie; which is six computed Miles from Dallribzian. Dallribzian from the Spittle of Glen Shee is also six Miles. The Water of Glen Shee and the Black Water are the same River.

Copied by F. Gould 12.t July 1781.

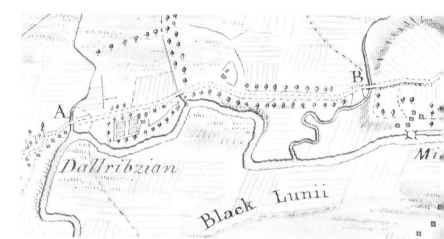

FIGURE 5.11

The Blairgowrie to Braemar road was eventually completed in 1750, when the six miles north of Dalrulzion, through the Spittal of Glenshee, was built (see detail from main map). It is perhaps not surprising that the Welsh Fusiliers and General Guise's Regiment failed to complete the road as they had 'rough' country and several passes to cross, through which 'rapid roaring Torrents full of Rocks' flowed. In contrast, Glen Shee was considered 'very pleasant, having a good Deal of Woods & several improved Spots' although the Black Water was 'as rough & rapid as those mention'd above'.

Source: George Morrison, *Survey of the Different Parts of the Road joyned between Blair Gowrie and Brae Mar; Made by a Detachment of one Hundred Men of Lord Viscount Bury's Regt* (1750).

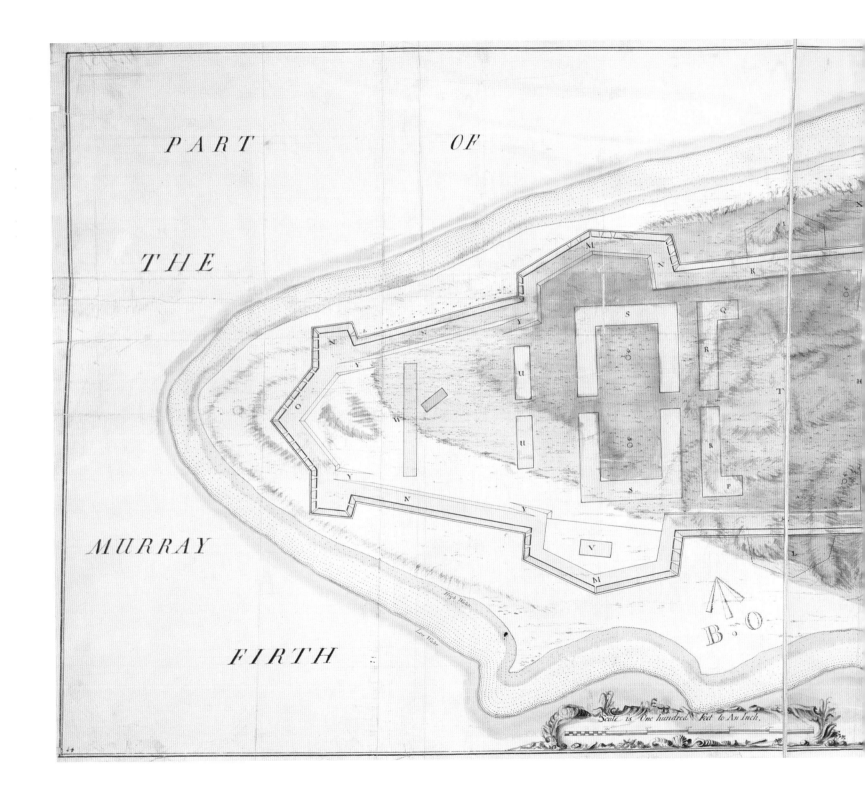

THE ROY MILITARY SURVEY TO FORT GEORGE, ARDERSIER

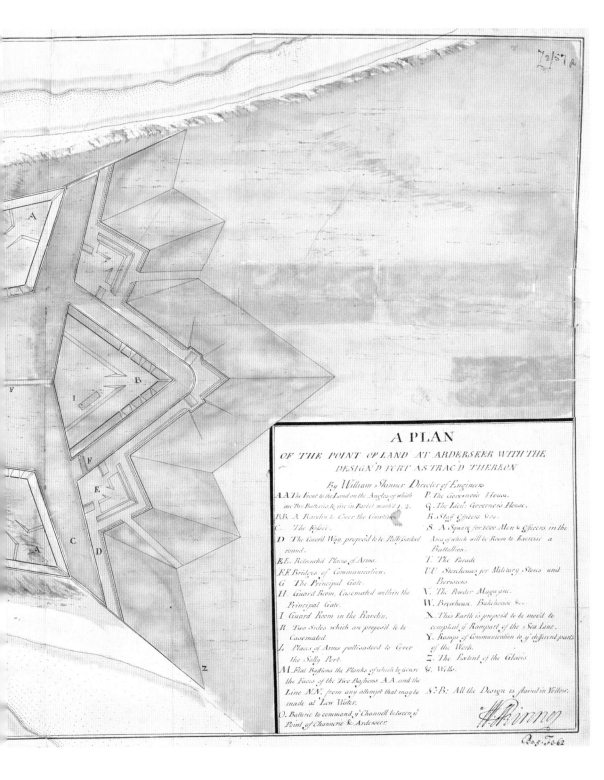

FIGURE 5.12

This slightly unusual plan shows the landscape of Ardersier peninsula on top of which is traced the ghost of a fortification 'staind in Yellow' following the cartographic colour convention of the time for a proposed building yet to be built. William Skinner selected and surveyed the site, 'the situation' in his view 'cannot be better chose; we have found Water, and that I think very good' – four wells are shown on the plan. A supply of fresh water was crucial to any garrison – a besieged place without sufficient water was destined to surrender. Skinner's only objection was the bleakness of the situation – a shingle spit jutting northwards into the Moray Firth, nine miles north-east of Inverness and five miles from Culloden – but 'good warm Barracks' would remedy this.

When the Duke of Cumberland saw the plans and estimate of costs for the new fort, he considered it 'a very sensible and Reasonable Scheme', one that should be 'put in Execution Immediately'. Skinner's first task was to purchase a circumference of two miles of land around the site, which would 'prevent making any Cover thereabouts, and it will give room for Gardening for the use of the Garrison'. The land, owned by John Campbell, a Lord Commissioner of the Treasury, was eventually bought by the Crown for £2,200.

Source: William Skinner and Charles Tarrant, *Plan of the Point of Land at Arderseer with the Design'd Fort as Trac'd Thereon* (1748).

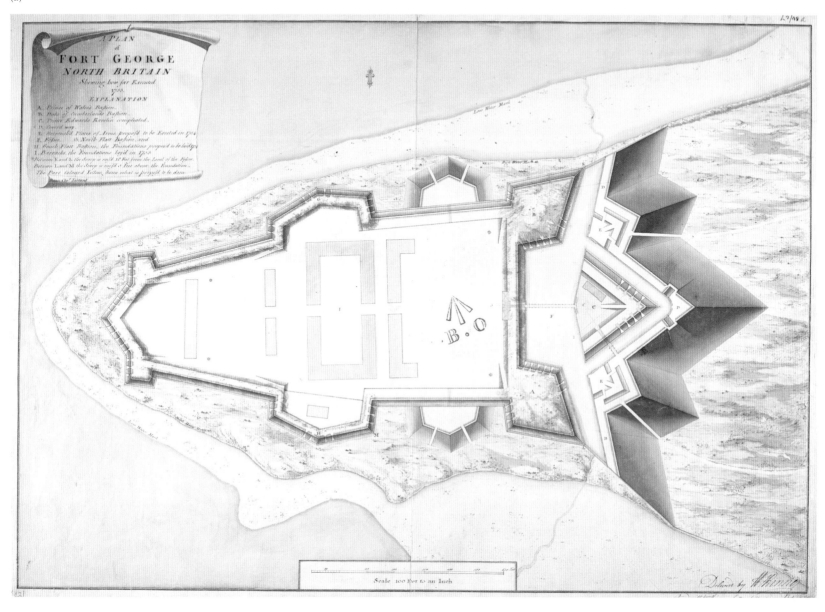

FIGURE 5.13

Fort George at Ardersier Point was a massive undertaking. In just over twenty years and at an eye-watering cost of £200,000 (around £20 million today), 'the most considerable fortress and the best situated in Great Britain' was built. The bastioned curtain walls of William Skinner's design follow the shape of the spit, culminating in a large sea battery at the point. The landward-facing defences feature two bastions, a ravelin (the external, triangular defence to the right) and a covered way with a ditch that could be flooded or drained via sluices running alongside. Within the outer defences were the expected military structures – the barracks, powder magazine, ordnance and provision stores, as well as a gun battery overlooking the Moray Firth. Building materials were transported by boat, so a harbour was one of the first structures to be built. The workforce numbered thousands – skilled workmen came from the Lowlands and Ireland, while the army, as with the road-building programme, provided the unskilled workmen. Barrel's Regiment, for example, worked on the construction in 1749.

Sources: (a) William Skinner and Charles Tarrant, *A Plan of Fort George, North Britain, Shewing how Far Executed* (1753). (b) William Skinner and Charles Tarrant, *Section thro' the Line 1.2; Section thro' the Line 3.4; Section thro' the Line 5.6.7; Section thro' the Line 8.9* (1753).

(b)
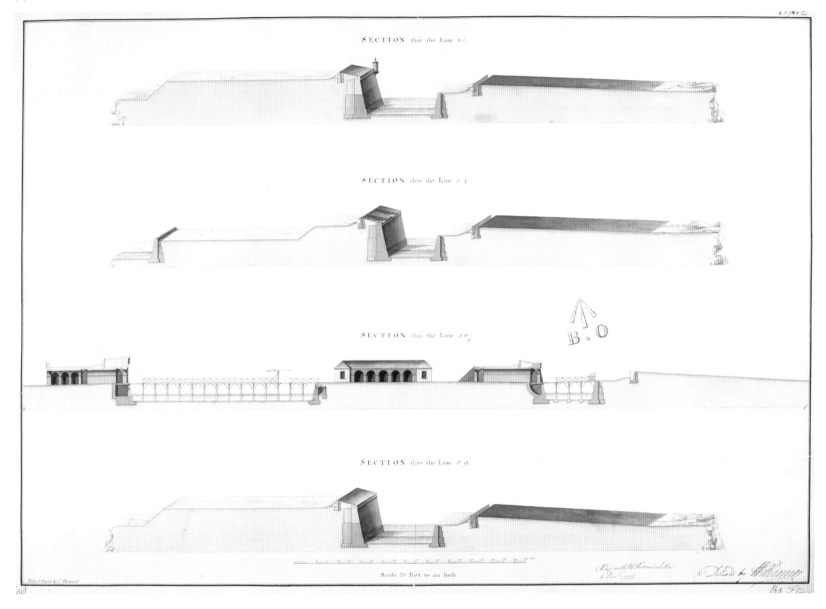

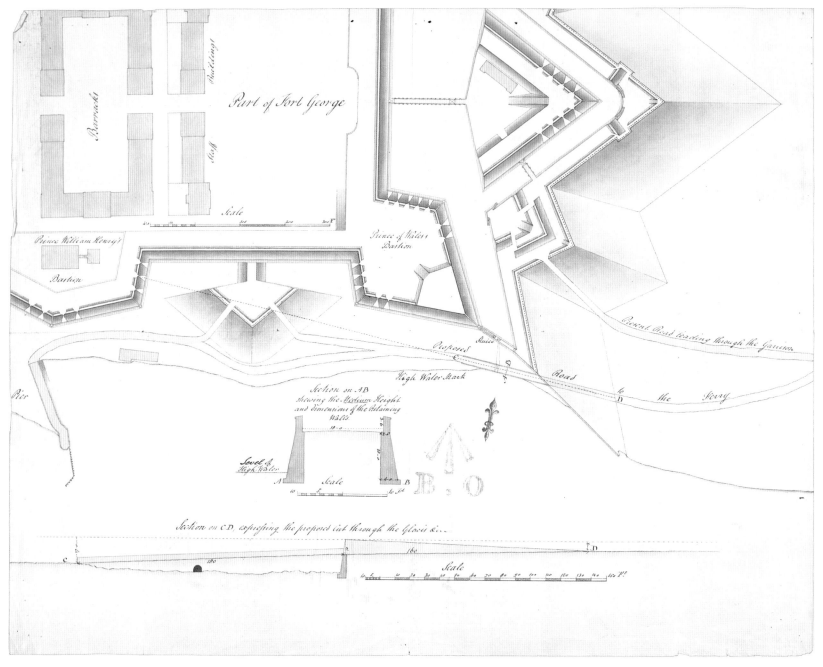

FIGURE 5.14

Military engineers often needed to turn their hand to practical subjects, such as re-routing roads and preventing them from flooding. Until 1787, the road leading to the pier at Fort George passed through the fort. Major Andrew Frazer, Chief Engineer for Scotland, proposed a new route that would skirt the fort's outer defences. As Frazer's plan shows, the new road would have to cross the high tide mark below the southern bastion, and so retaining walls would need to be built to prevent the road from flooding.

Source: Andrew Frazer, *Present Road leading Through the Garrison [and] Proposed Road to the Ferry* (1787).

at Anstruther before finally abandoning his plans. In response, a battery was rapidly built to cover the entrance to Leith harbour. A more substantial 'Inclosed Battery or Redoubt' was then built in 1780, with elevations designed by James Craig, the architect of Edinburgh's New Town (fig. 5.15).

Despite the growing threats to Scottish harbours of privateers, especially from the late 1770s when France and then Spain joined the American colonial states in fighting against Britain, the Board of Ordnance dragged its feet over defensive works for financial reasons. To save money, they encouraged coastal towns and small ports to construct their own defensive works. To assist in this, smaller-scale charts of sections of coastline were produced to plan the location of batteries to protect shipping (fig. 5.16). Following orders in 1781 that 'Fort Charlotte should be put in a complete state of defence', the fort at Lerwick on Shetland was repaired and re-equipped to protect the Sound of Bressay, which prompted detailed charting of that coastline as well as designs of the fort, its barracks and powder magazine (fig 5.17).

Arguably, the most impressive Scottish fortifications and maps of all time were made during this period – the whole Scottish mainland comprehensively mapped in less than eight years, and the construction of the monumental Fort George at Ardersier, the most impressive modern artillery fortification in Scotland. With the benefit of hindsight, some may criticise them both for being overblown military exercises, whose real military purpose was questionable before they were even completed. Ardersier was never to see any hostile artillery fire, and the Roy Military Survey was hardly consulted after its drafting, only used a half-century later to help construct a topographic base for Aaron Arrowsmith's civilian *Map of Scotland . . .* (1807). These are fair points, but they are not the whole story. The Roy Military Survey and Fort George both used leading expertise, techniques and a visionary sense of purpose to create something genuinely new and impressive, quite apart from their unrivalled aesthetic and utilitarian qualities at the time and for later ages. Huge labour was collectively expended in their creation, and the results still continue to impress us today. Both provide windows into the military landscape of the mid eighteenth century and how it was perceived – not merely what was there then, but why it was there, who created it and needed it, and why it mattered. They were both enduring statements of Hanoverian authority and power, in the landscape and on paper, and both played an important role in diminishing the Jacobite threat. In the history of Scottish fortifications and maps they stand apart, and they were to have an enduring and significant influence on later fortifications and maps, a theme we take up in the next two chapters.

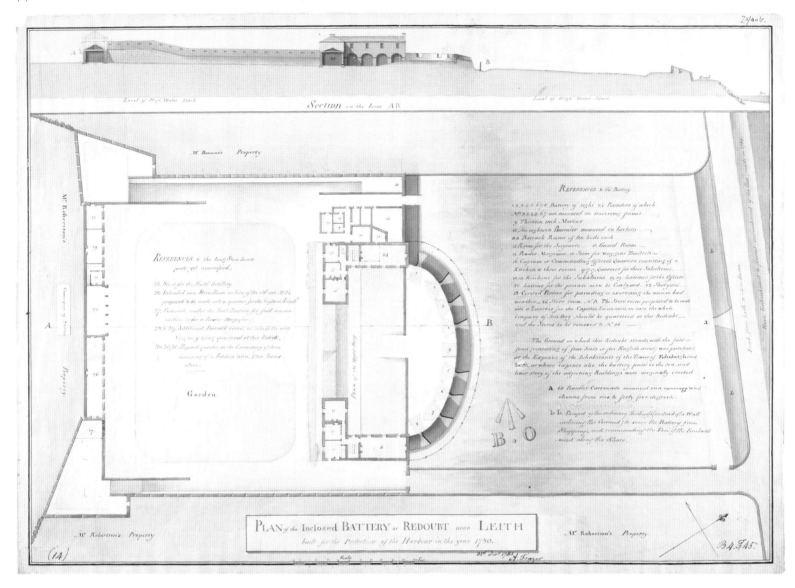

FIGURE 5.15

(a) Leith Fort was constructed from 1780 following the near attack on Edinburgh by Privateer John Paul Jones in September 1779. The initial response by the Edinburgh and Leith Town Council was to throw up a battery of nine guns covering the entrance to Leith harbour, but a more substantial 'Inclosed Battery or Redoubt' was then built in 1780. This structure is shown as the central part of this plan and elevation drawn in December 1785 by Andrew Frazer, who had recently been engaged on significant reconstruction work at Fort Charlotte in Lerwick, Shetland (fig. 5.17). Frazer was an expert draughtsman, who also usefully wrote very precise and detailed notes on his plans. These confirm that the original construction had been funded by the Edinburgh and Leith Town Councils, but was then taken over by the Board of Ordnance, with plans to construct two substantial bastions (shown on the left of the plan) to the rear (south-west) of the main battery. The buildings connecting the bastions were to house artillery, stores and barracks. The main artillery defences to the front consisted of a curved battery of eight 24-pounder guns, while no. 10, at the top of the plan, was 'an 18 Pounder mounted *en barbette*' (on a raised platform), shooting over the parapet. Frazer even shows the 'Covered Portico for Parading or Exercising the Men in Bad Weather' (no. 23).

Although significant building work and repairs were recorded at the fort – thirteen separate substantial episodes between 1795 and 1815, costing a total of £49,203 – the bastions to the south were not constructed, and the fort never saw military action. The threat of a French invasion

(b)

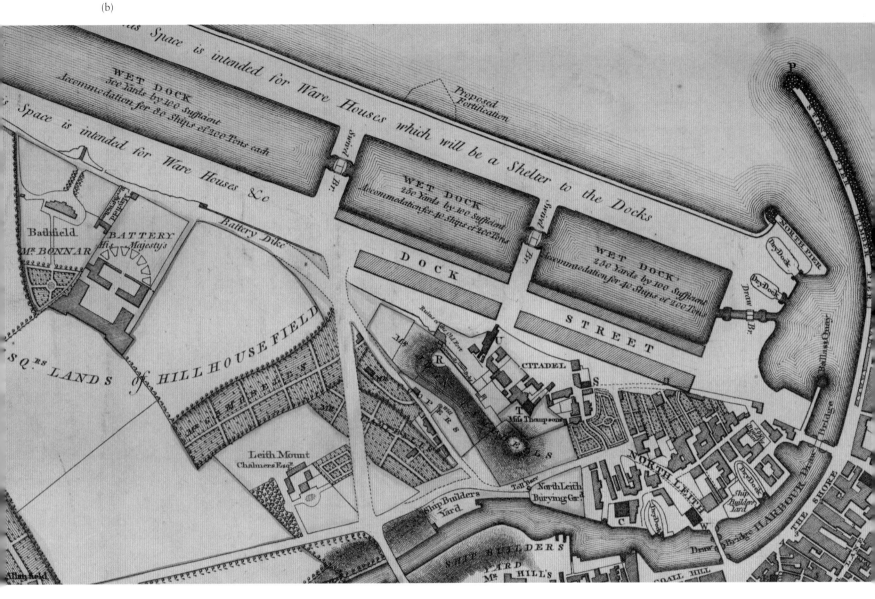

did not materialise, and the construction of the large wet docks, to the west of the harbour and directly in front of the fort made its defensive role redundant. In 1824, the artillerymen in the fort assisted in the 'Great Fire of Edinburgh', hauling their fire engine up to the High Street, and from 1861 they lent their gun to Edinburgh Castle's Half Moon Battery to fire the famous 'One O'clock Gun'. However, the main role of the fort in the nineteenth century was as an artillery depot, gunpowder store and barracks. Following local protests, Blackness Castle became the main gunpowder store in Scotland from the 1870s, and its Coastal Artillery School moved to Broughty Castle in 1908. The fort fell out of military use in 1956 and its interior was subsequently demolished, with further development in 2013 erasing the peripheral remains.

(b) While Frazer's military plan is useful in giving a clear distinction between what was existing and what was planned, John Ainslie's much more well-known 1804 plan blends reality with anticipated plans, showing the fort with its planned bastions to the south, along with the four wet docks (of which only the two to the east were ever constructed).

Sources: (a) Andrew Frazer, *Plan of the Inclosed Battery or Redoubt near Leith, built for the protection of the Harbour in the Year 1780* (1785). (b) John Ainslie, *Old and New Town of Edinburgh and Leith with the proposed docks* (1804).

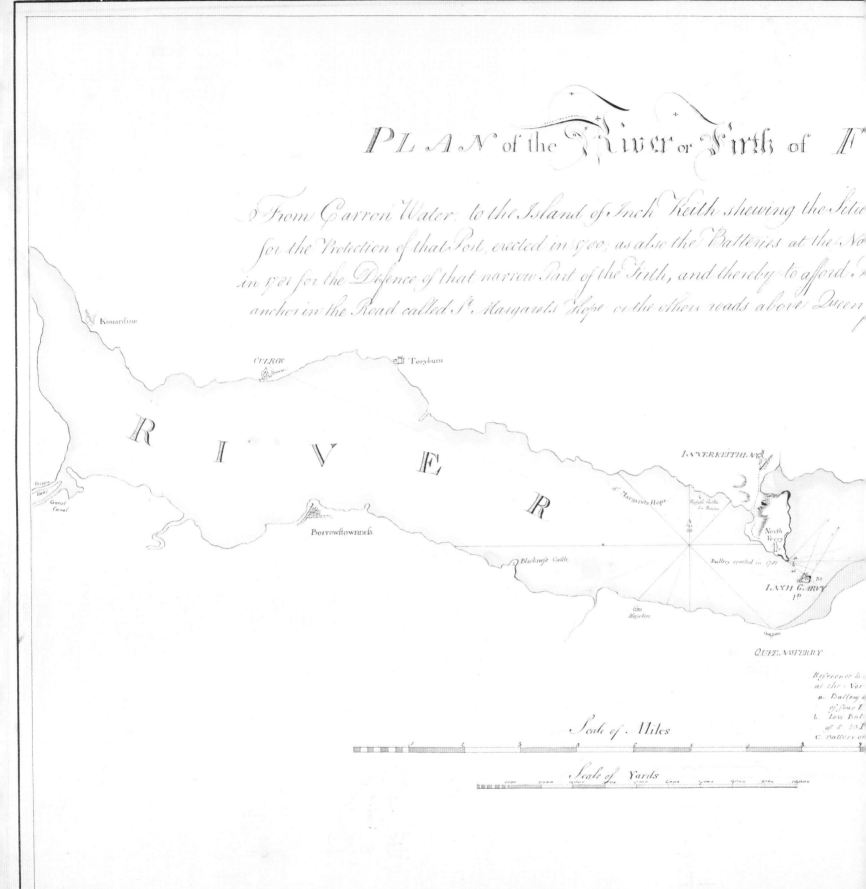

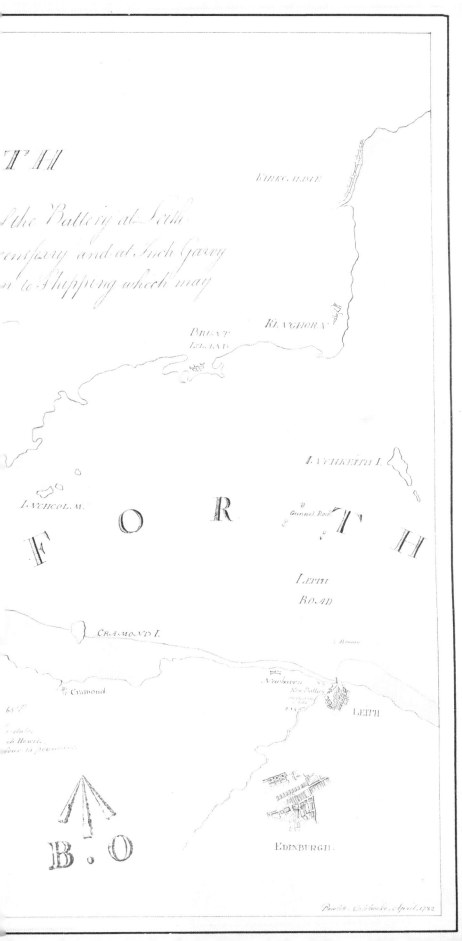

FIGURE 5.16

This plan extends from Kincardine in the west to the island of Inchkeith in the east, and shows the towns situated along the shore of the Firth of Forth, as well as several batteries: at Leith 'for the Protection of that Port', built in 1780; at North Queensferry and Inchgarvie (see detail from main map), built in 1781 'for the Defence of that narrow Part of the Firth, and thereby to afford Protection to Shipping which may anchor in the Road called St. Margarets Hope or the other roads above Queensferry'.

Reference to the batteries at North Queensferry:
    a. Battery on the height of 65 ft of four Field Pieces.
    b. Low Battery 12 ft above the Water of 8,
       20 Pounders & 1.8 Inch Howitz.
    c. Battery on the Island of four 14 Pounders.

The map was drawn for the Duke of Richmond, appointed Master-General of the Ordnance in 1782. When reviewing the defence of Scotland in 1797, War Secretary Henry Dundas remarked that the batteries would 'countenance against any small predatory Landings, which may be attempted on any of the different extended Coasts of Scotland, and against which if they escape the vigilance of our Navy'.

Source: Pawlett William Colebrooke, *Plan of the River or Firth of Forth* (1782).

(a)

Nº 1.

Nº 1. PLAN OF FORT-CHARLOTTE in March 1781.

Section from A to B in March 1781.

Section from A to B January 1783.

FIGURE 5.17

(a) Fort Charlotte overlooks the important natural harbour of Bressay Sound in Shetland. The fort was originally planned as a rough pentagon with corner bastions by John Mylne, Master Mason of the Crown from 1636, during the Cromwellian invasion of Scotland. There were many delays in building the fort in Lerwick due to a lack of suitable stone, lime and workmen, although the reopening of hostilities against the Dutch in 1665 provided a renewed impetus for construction. The fort successfully held off a Dutch fleet in 1667, who thought it more heavily armed than it really was, despite its walls being unfinished. Work was called off and the garrison was disbanded following the end of the Second Anglo–Dutch War in 1667, a decision which proved premature. With the resumption of war against the Dutch during the Third Dutch War (1672–77), the Dutch attacked again and burnt the barrack block, along with several other buildings in Lerwick. The fort lay abandoned and incomplete for over a century until concerns over American privateers in the 1780s resulted in a decision to rebuild it along similar lines, naming it Fort Charlotte in honour of George III's queen. At this time, Andrew Frazer created a set of detailed plans to show the planned reconstruction of the fort, barracks and environs. This plan and set of profiles depicting the state of the fort before work commenced are our best indication of the original seventeenth-century fort.

(b) Frazer's plan of Bressay Sound shows the fort after its reconstruction in the eighteenth century. The entrance to the bay was well protected. As noted on Frazer's plan, ships entering the bay could be seen for '6 or 7 leagues' from South Brae Head – a high hill '140 feet above High Water' commanding the southern entrance and the landing places in the bay. Redoubts were begun but never completed on both South and North Brae Head. Instead, a small battery was built to cover the landing places to the north-east of the fort.

Sources: (a) Andrew Frazer, *Plan of Fort Charlotte in March 1781* (1781). (b) Andrew Frazer, *Plan of the Bay called Brassa Sound* (1786).

FIGURE 5.18

(*Overleaf*) George Morrison, *Survey of the Different Parts of the Road joined between Blair Gowrie and Brae Mar; Made by a Detachment of one Hundred Men of Lord Viscount Bury's Regt* (1750).

(b)

CHAPTER SIX
THE FRENCH REVOLUTIONARY WARS TO THE FIRST WORLD WAR, 1792–1918

The French Revolutionary and Napoleonic Wars (1792–1815) and First World War (1914–18) provide convenient start and end points for a long century of military mapping, dominated by these and other conflicts. The threat of coastal attacks to Scotland grew, while the Crimean War (1853–56) and the Franco–Prussian War (1870–71) also had important military consequences in Britain. This period witnessed a major transition in the types of fortification and their geography, with a decline in the role of major inland fortifications and a growth in small, scattered coastal batteries, especially on the east coast. There were also major changes in the scale of armies, in the power of artillery and guns, and through new steam power and rail transport. By the First World War, there were also new types of weapon to contend with in the form of the submarine and torpedo. The British institutions responsible for defence and mapping changed following a major expansion of state-funded map-making in the nineteenth century, and the style and aesthetic of military maps was completely transformed by the growth of lithography as a printing technique.

Coastal batteries and martellos

The renewed declaration of war by France against Britain in 1793 created serious worries of a French invasion, even if the threat ebbed and flowed with France's other military preoccupations. The Board of Ordnance made a comprehensive report on the state of British defences early in 1794, generally confirming that only those permanent defences in Plymouth, Portsmouth and Chatham were thought to be in any way adequate. The chain of forts and batteries that stretched along the coasts were too far apart to be of any real use, and many of them were in serious disrepair, or were disarmed and unmanned. There was a French landing in Wales in 1797, and a more serious attempt in Ireland in 1796–97, showing that the threat was all too real. With only a small standing army, militias and part-time volunteer corps were hastily assembled, with men raised for fencible regiments in Scotland. Defending the whole coast was impractical, so the decision was made to station troops in likely landing places, and batteries were built in almost every harbour (fig. 6.1). Dumbarton Castle, commanding the Clyde, was well armed, while in the Firth of Forth, guns were mounted at Blackness, Leith, Queensferry, Inchcolm and Inchgarvie. The French also prepared military maps of Scottish coasts, based on the best available Scottish charts (fig. 6.2).

From 1803, with the obvious presence of a French invasion army across the Channel, more defences were needed. These

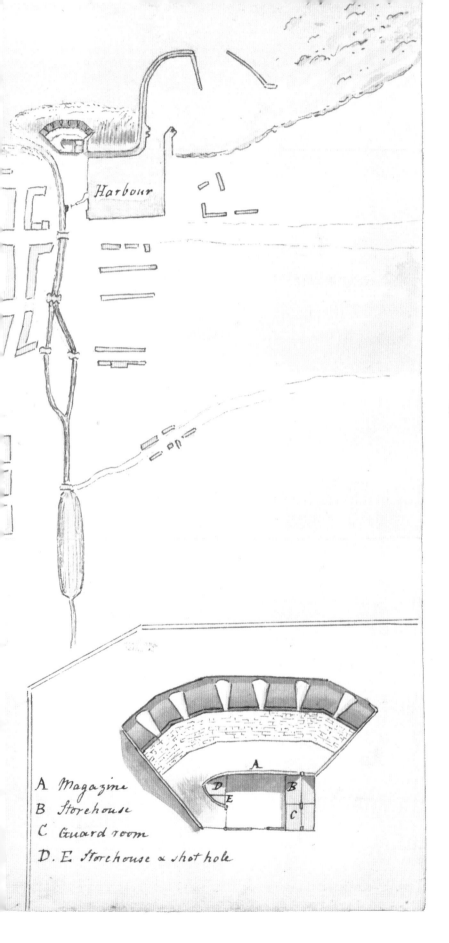

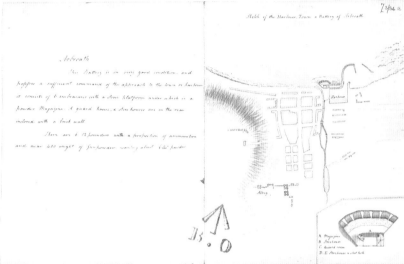

FIGURE 6.1

In 1785 William Gravatt, a Board of Ordnance engineer, was ordered to visit and review the main coastal harbour defences in Scotland which had been constructed by local towns. He made hasty sketches with a brief report of several batteries, including Arbroath shown here, as well as those at Campbeltown, Greenock, Aberdeen, Montrose and Dunbar.

Four years earlier, in May 1781, Arbroath had been fired on by Captain Fall, the commander of a French privateer. As was customary, he demanded money (some £30,000) and took hostages (six unlucky local residents). General alarm followed in the town, some townsmen gathered on the beach with firearms, but Fall then decided that plundering various sloops nearby seemed a more profitable course of action, and the town was narrowly spared.

As Gravatt's sketch shows, the Town Council hastily put up a battery soon afterwards as a means of defence. Positioned immediately to the east of the harbour (see detail from main map), its basic design has much in common with the contemporary batteries elsewhere along the coast. As Gravatt wrote, 'This Battery is in very good condition, and . . . is a sufficient command of the approach to the town or harbour . . . of 6 embrasures, with a stone platform under which is a Powder Magazine.' There were six 12-pounder guns and a good supply of gunpowder, which it is hoped offered some reassurance to the locals, although it does not seem that the guns were ever fired in anger. The battery was dismantled soon after the cessation of hostilities in 1815. The sketch map serves its purpose well in picking out the essential location and detail of the battery, along with the key features of the town in the immediate context.

Source: William Gravatt?, *Sketch of the Harbour, Town & Battery of Arbroath* (c.1795).

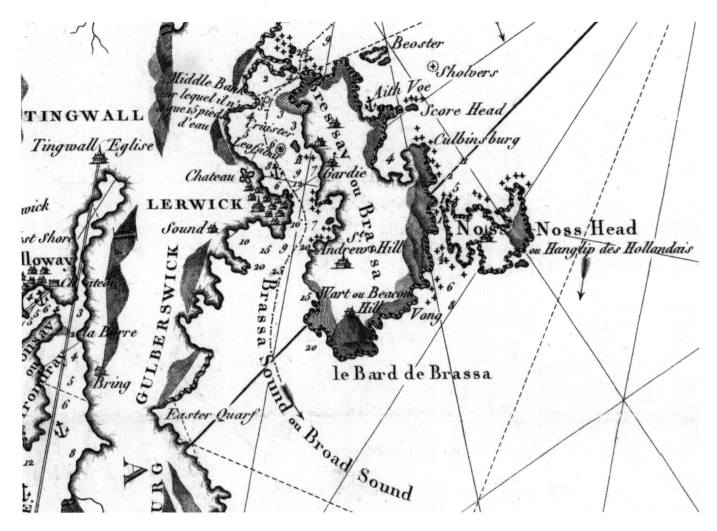

FIGURE 6.2

In order to plan for a likely invasion of Britain, the French Dépôt Général de la Marine gathered the best available British charts they could find, re-engraving them with suitably translated detail. For Shetland, the most recent British Admiralty survey by Lieutenant Edward Henry Columbine in 1795 was not generally available outside Admiralty circles until its use in Arrowsmith's *Map of Scotland* . . . (1807), and so the French Dépôt used the work of 'Capn Anglais Preston' from the 1740s, revised by Paul de Loewenorn in 1787. Thomas Preston was an English mariner in the tobacco trade, whose ship was wrecked in the region of Skelda Voe in January 1743 while he was *en route* from Virginia to London. He was detained in Shetland until May 1744, but spent his time usefully constructing a detailed chart of the islands with extensive notes, anchorages, tidal stream information and outlines of the hills as seen from the sea. Preston's chart was improved upon, especially in terms of its positional accuracy, by Paul de Loewenorn, a Danish naval officer, who had served with the French in the American Revolutionary War and spent time in Shetland's waters in 1786. The chart, therefore, has an interesting provenance through English, Danish and French hands, with a combination of Old Norse, English, Dutch and French captions and place names – see, for example, 'Noss Head, ou Hanglip des Hollandais'. The recently reconstructed Fort Charlotte in Lerwick is clearly marked as 'Chateau', while further north the chart faithfully translates details such as Preston's warning of 'Middle Bank on which is but 15 feet of water'. The chart was published in Year XIII of the new French Revolutionary Calendar, dating from the foundation of the Republic in 1793, which would be abolished the following year, L'An XIV (1805).

Source: Dépôt Général de la Marine, *Carte des Îles Shetland* . . . (1804–05).

took the form of 'martellos' or independent gun towers. The name derives from Cape Martella in Corsica, where a British naval squadron in 1794 had failed to blast a signal tower into submission from a range of only 150 yards. The towers were vigorously promoted by Sir David Dundas, who had trained as an engineer under William Roy on the Military Survey of Scotland, commanded in Corsica in 1794, and later went on to become Commander in Chief of British Forces from 1809 to 1811. In the corrupted form of the name, they rapidly sprang up along the southern coast of England, where over 100 martello towers were built between 1804 and 1812. At first glance, they seem to represent a significant break from the large bastioned fortifications of the eighteenth century, but in practice, they were a temporary defensive work in response to a potential sea invasion. Large fortifications, albeit with modified bastions and other features, were to continue in popularity through to the early twentieth century. Three martello towers were also constructed at this time in Scotland: one on an islet off Leith harbour in 1810 and two in Orkney in 1813–15. The Leith martello was designed for three guns, and had the familiar tapering profile. The two Orkney martellos were intended to protect the battery at Hackness and guard the entrance to Longhope Sound (fig. 6.3). They were not built to resist invasion but to protect convoys of ships involved in trade with the Baltic from American privateers. As for the martellos, the batteries erected at this time were mostly of a simple and temporary nature, forming the segment of a circle in plan. Their guns were often mounted on platforms protected by a rampart of earth, which they fired over (fig. 5.15, fig. 6.1).

Infrastructure and prisoners of war

Following the decline of the Jacobite threat from the mid eighteenth century, the main state castles in Scotland had a steadily reducing defensive role to play in subsequent decades. Their roles shifted more to raising troops to serve in combat overseas, in Britain's expanding empire. However, with the drive to raise regiments in the 1790s, prompted by the Napoleonic Wars, there was a need to upgrade infrastructure and provide better accommodation for soldiers at all the major castles in Scotland. Within Stirling Castle, for example, new floors were put into James IV's magnificent Great Hall, constructed in the early sixteenth century, and the hammer-beam roof was removed for the creation of a third floor of barracks around 1800. Edinburgh Castle also saw major internal reorganisation, with the construction of the enormous New Barracks in 1796–99, which could house an entire infantry regiment (600 men). In these changed circumstances, military engineers were able to put their broad range of skills, covering subjects like hydraulics, design, construction and draughtsmanship, to requirements such as water supply for castles, a problem that had become particularly acute at Edinburgh (fig. 6.4).

A related and serious difficulty was the need to hold the steadily growing numbers of prisoners of war, which grew to number 13,000–14,000 prisoners in Scotland in the closing years of the Napoleonic Wars. During the Anglo–French War (1778–83) not only home waters, but also those around overseas imperial outposts became theatres of war, and many French boats and their crews were captured.

In July 1781, there were almost a thousand prisoners of war in Edinburgh Castle, including 140 desperate invalids from a French frigate, the *Marquis de la Fayette*, which had been sunk by the Royal Navy in the West Indies. The French seamen endured a harrowing fourteen-week voyage on scant rations, and around twenty-one later perished in Edinburgh Castle. These and other prisoners were mostly held in the fourteenth- and fifteenth-century castle vaults under the Great Hall and Queen Anne Building (fig. 6.5), although owing to overcrowding, the recently completed New Barracks (marked 29 on fig. 6.4), a short distance to the north-west of the castle vaults, were altered so that they could accommodate prisoners.

The responsibility for prisoners of war lay with the Transport Board, part of the Admiralty, until it was abolished in 1817, but it had growing difficulties of where to house prisoners. Edinburgh City Bridewell (the jail) was used in

(a)

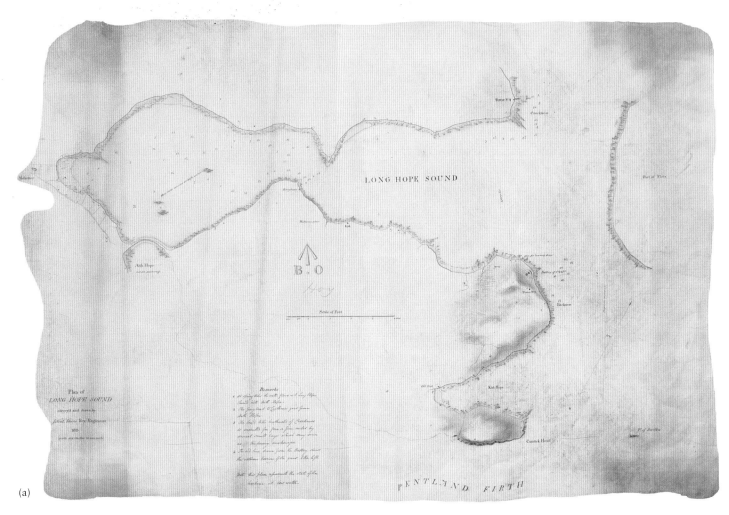

(b)

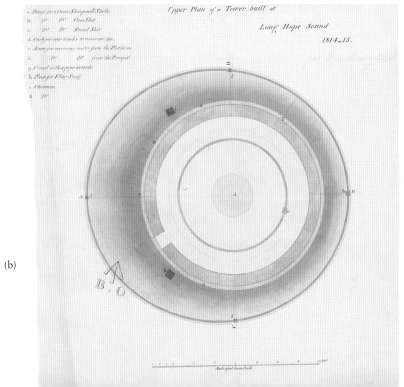

FIGURE 6.3

(a) The Hackness martello tower in Orkney, along with its associated battery 180 metres to the north-west, was constructed in 1813–15 to protect the eastern entrance to the defended anchorage of Longhope, situated between the islands of Hoy and South Walls near the southern entrance to Scapa Flow. Lieutenant Philip Skene of the Royal Engineers surveyed and compiled a detailed plan of Longhope Sound, including rock outcrops and soundings, showing the position of the tower and battery, as well as the tower at Crockness, on the opposite side of the water. Following the declaration of war by the United States in 1812, the towers were intended to protect convoys of ships in the Baltic trade from American privateers.

(b) As indicated in the *Upper Plan of the Tower*, the wall on the seaward side was twice as thick in order to withstand bombardment, creating an elliptical plan. The interior floors were circular, and stairs within the thickness of the wall led down to a storeroom and magazine at ground level and up to a parapet and gun platform at the top.

(c) The tower was around 10 metres high above the ground, with an excellent view over the approaches to the Longhope anchorage, and the 24-pounder cannon mounted there could also protect the landward side of the battery from attack. The tower had its own water supply from a cistern built into the foundations, and water could be raised to the living quarters using a hand-pump.

The wedge-shaped battery nearby had eight 24-pounder guns, mounted on two faces, with barracks, guardhouse and magazine symmetrically arranged to the rear and partially built underground. The guns swept the south-east approaches to Longhope through Switha Sound and Cantick Sound. In 1866, the defences on the tower and battery were upgraded to allow heavier guns and protection, following concern about the possibility of another French invasion, and the tower was also used as a naval signal-post during the First World War.

Sources: (a) Philip Skene, *Plan of Long Hope Sound* (1815). (b) Philip Skene/Robert Hoddle, *Upper Plan of the Tower Built at Long Hope Sound* (1815). (c) Philip Skene/Robert Hoddle, *Section through C.D. of the Tower Built at Long Hope Sound* (1815). All images courtesy of The National Archives.

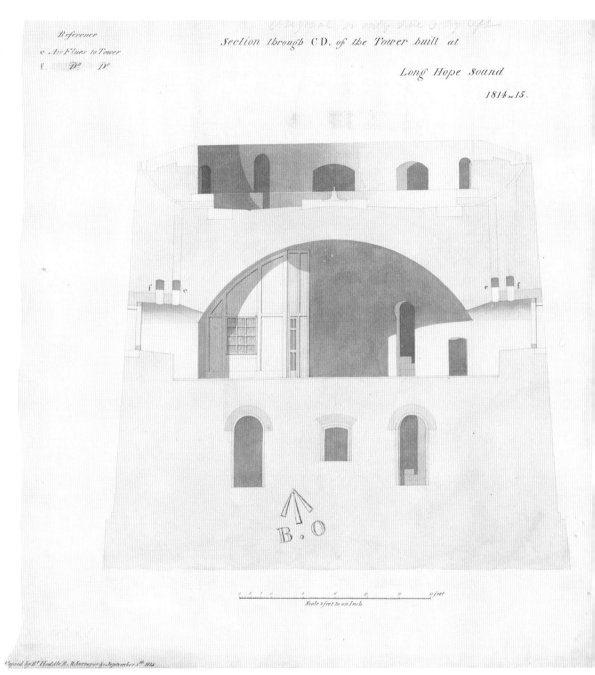

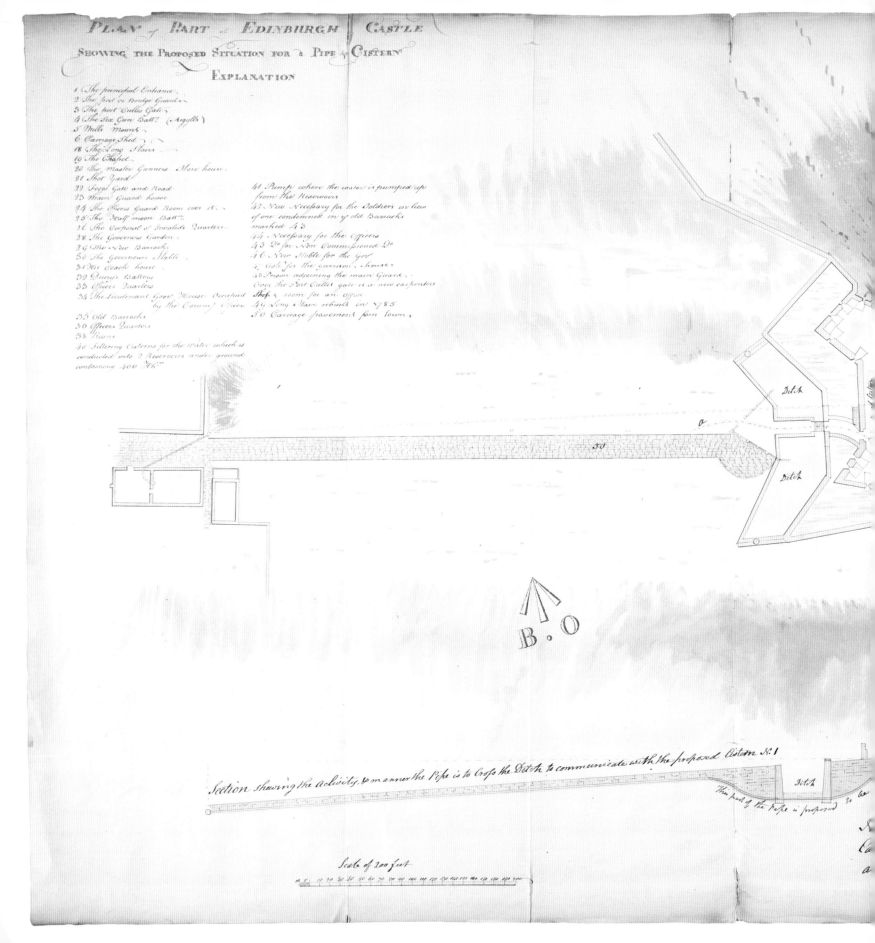

FIGURE 6.4

A good supply of water was essential for any garrison, although for centuries Edinburgh Castle struggled to get a decent water supply.
In fact, during the siege of 1571–73, the blocking of the Forewell, positioned in the spur (fig. 2.4), was an important factor in the ultimate surrender of the castle. Although this well was subsequently restored, and a Backwell to the west of the castle was dug in 1628, the castle increasingly became dependent on the town reservoir, which had been piped to Castle Hill from 1676. This was clearly a problem and, as General Bland wrote to William Skinner from the castle in 1755, 'what is most wanted in the Castle at present, is water'. For centuries, European military engineers had included hydraulics as part of their education, and this plan illustrates this well.

Henry Rudyard's plan of 1794, orientated with north to the bottom, shows the initial solution to the problem – a plan to construct pipes made of elm from the reservoir at the head of the High Street on the left, to new cisterns just behind the outer gate, beneath the Half-Moon Battery. The long curved route, shown in the lower half of the detail from the main map, and marked a, b, c, d, e, shows 'the distance the Troops stationed in the Castle have to go to the Reservoir belonging to the City for Water'. The plan was soon implemented, with a filtering cistern (40) and pump (41) also put in just inside the Half-Moon Battery. The plan also gives a rare glimpse into the garrison's sanitary arrangements, which were given a substantial upgrade at the time. The building marked '42' on the plan, just to the north of the Officers' Quarters (36), was the 'New necessary for the Soldiers in lieu of one condemned in ye Old Barracks marked 43'. 43 can just be seen as the projection over the parapet wall beyond the Old Barracks (35) at the top of the plan. Segregated toilets were also installed for the higher ranks. Between the New Barracks (29) and the Chapel (19) was a new 'Necessary for the Officers' (44) and 'Ditto for Non Commissioned Officers' (45). The new cistern was 16 by 12 feet in dimensions and held 7,200 gallons of water.

Source: Henry Rudyard, *Plan of Part of Edinburgh Castle Showing the Proposed Situation for a Pipe & Cistern* (1794).

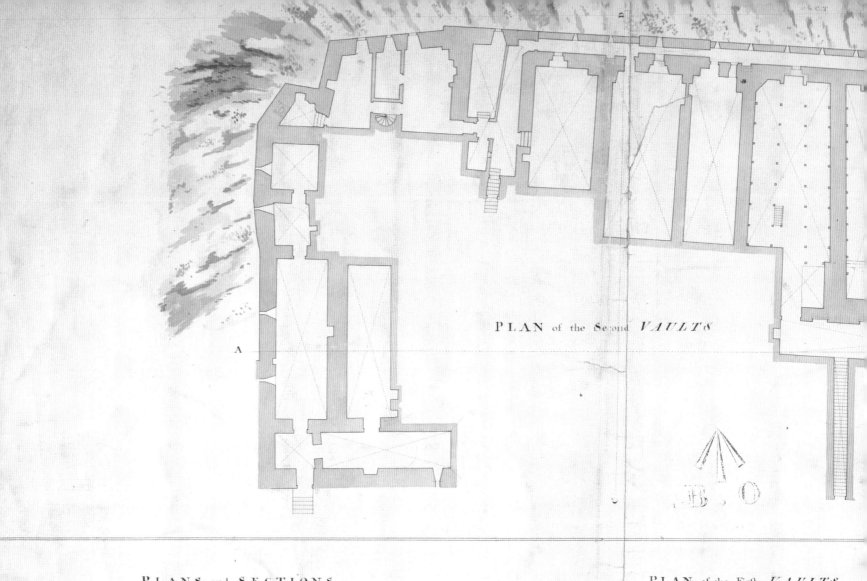

PLAN of the Second VAULTS

PLAN of the First VAULTS

Prisons for Soldiers

PLANS and SECTIONS
of the Several
VAULTS and FLOORS
of the
GOVERNORS and LIEU:T GOV:RS HOUSE.
the FORT MAJORS. SUBALTERN OFFICERS
of the INVALIDS.
and CHAPLAINS HOUSE.
the SOLDIERS BARRACKS.
and the ARSENAL.
in EDINBURGH CASTLE.
1754.
Shewing to what use they are appropriated.

Scale 14 Feet to an Inch

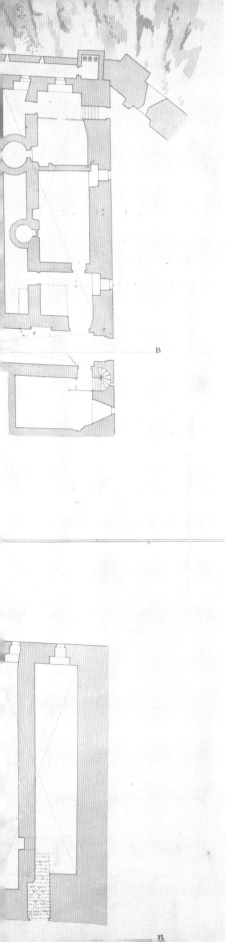

FIGURE 6.5

This plan, dating from 1754, shows the two tiers of cavernous stone vaults beneath the Great Hall and Queen Anne Building in Edinburgh Castle. The vaults date from the fourteenth and fifteenth centuries, and initially were used for stores of food and military provisions. During the eighteenth century, and especially during the Napoleonic Wars, the vaults housed prisoners of war. On the lower half of the plan, showing the lower vault, the two rooms on the left are marked as 'Prisons for soldiers' in 1754. In fact, it is likely that these two rooms and the ones directly above them were originally built as prisons. As shown on the plan, they had only a tiny, narrow ventilation slit, and a latrine at the top of a steep stair. All four were accessed from the 'Devil's Elbow' walkway, which led from the east side of the Great Hall. Jacobites are known to have been imprisoned here. In 1757, during the Seven Years' War, the 78-strong crew of a French privateer, the *Chevalier*, were incarcerated in the vaults, probably in the second vault from the left, at the top. The names of French crewmen are carved into the door here, and there is also graffiti, including pictures of sailing boats. By 1759, 362 prisoners of war were here, but around 500 were repatriated at the end of hostilities in 1763.

During the American Revolutionary War (1775–83), and particularly during the French Revolutionary and Napoleonic Wars (1792–1815), the vaults' role as prisons peaked. Many of the prisoners were sailors, but soldiers later arrived, particularly from Spain and Portugal. In 1799, there were 750 prisoners held in the vaults even though their maximum capacity was supposed to be 450. The increasing number of countries involved in the conflict also resulted in a growing diversity of different nationalities and backgrounds of prisoners. Tensions rose in the squalid and overcrowded conditions. The French and Spanish 'constantly engaged in feuds' and had to be separated, while friction between the French and the Danes finally resulted in the latter being sent south to Portsmouth.

Escape was a high priority, and several managed it in ingenious ways, although many more perished in the attempt. The mass breakout of 49 prisoners in April 1811, by cutting a hole in the south parapet wall, finally sealed the fate of the castle's role as a prisoner-of-war depot. Although these escapees were eventually recaptured – apart from the poor fellow who died when he lost his hold on the castle rock and fell – the Admiralty declared in December that 'in consequence of the ease with which French prisoners-of-war can escape from Edinburgh Castle [we are] pleased to direct that this depot be abolished'. By 1814, the last remaining French prisoners were marched out and the vaults were handed back to the garrison. They were described as being in a 'filthy and dirty condition', but were soon put to use again as stores.

The vaults were used for prisoners again – primarily German – during the First World War; the deepest, westernmost prison was reserved for 'traitors', including David Kirkwood (1872–1955), who was imprisoned in 1916 for his part in the munitions workers' strike at Beardmore's Parkhead Forge in Glasgow. Kirkwood was later released following the personal intervention of Winston Churchill.

Source: Charles Tarrant, *Plans and Sections of the Several Vaults . . . in Edinburgh Castle; Shewing to what Use They are Appropriated* (1754).

1803–04, but several prisoners managed to escape. In 1804, Robert Trotter of Castlelaw finally agreed to lease the old mansion house at Greenlaw, north of Penicuik. Greenlaw was demolished after the Napoleonic Wars and rebuilt as Glencorse; it survives as an army barracks today. Further accommodation was found nearby in Esk Mills and Valleyfield Mills from 1811. The Perth depot, the only purpose-built depot for prisoners of war in Scotland during the Napoleonic Wars, and subsequently part of Perth prison, took further overspill. After Edinburgh Castle therefore, Glencorse Barracks and Perth prison (although greatly altered and reconstructed) are the largest remaining visible legacies today of prisoner-of-war camps from the Napoleonic Wars in Scotland; maps provide a key insight into their use at the time.

From the end of the Napoleonic Wars there was also a significant change in the method of drafting military plans; the maps made at this time were some of the last hand-drawn military plans by the Board of Ordnance in Scotland and elsewhere. Following the invention of lithography by Senefelder in the late eighteenth century, lithography was used for military map-making purposes before and after the Peninsular War (1807–14). Lithography depends on the mutual antipathy of water and grease. A map is drawn on a porous stone or metal surface using a grease-based liquid. An oil-based ink is then applied to the moistened stone or metal; the ink is repelled by the moisture but sticks to the grease-treated drawing. When a sheet of paper is laid over the stone/metal and pressed down, the map image is printed onto the paper. The process was relatively simple, cheap and quick, especially compared with the hand copying of maps, as favoured in the past by the Ordnance Drawing Room for training purposes. Lithography paved the way for the mass production of cheap maps for civilian and military audiences in the nineteenth and twentieth centuries. In addition, the development of photo-lithography (often called photo-zincography by Ordnance Survey) in the later nineteenth century allowed quick and easy copying of a base map, for subsequent overprinting of military information, which became a very common form of military map in the twentieth century.

Internal pacification and riot control

In the decade after the end of the Napoleonic Wars in Scotland, military infrastructure and defensive works were constructed particularly with an eye to internal pacification and riot control. While there were food shortages and meal mobs in Scotland in the 1790s, there was none of the mass mobilisation of dissent that characterised France and Ireland. For the next decade, the French blockade and plans for invasion were more likely to promote patriotism than rebellion. Nevertheless, there was an important spate of barrack-building in the 1790s, with some maps showing barrack plans drafted by the Board of Ordnance, and with barracks increasingly appearing on civilian maps (fig. 6.6). The pressure to construct new barracks grew rapidly after the end of the Napoleonic Wars. A profound economic depression in 1816 with high food prices coincided with longer-term falls in real income; between 1816 and 1831, Glasgow weavers saw their income fall by nearly one-third. The demobilisation of soldiers meant that many of the disaffected, radicalised groups were well experienced in the use of weapons, and reports of plots to violently overthrow the government in 1817 seemed very real. The Peterloo Massacre in Manchester in August 1819, and the ruthless suppression of the 'Radical War' in south-west Scotland in April 1820, were only part of a wider systematic government repression. Both saw the use of cavalry to charge on the crowd. The Radical War had brought together around 60,000 people, particularly in Glasgow and nearby towns, who proclaimed the unity of all classes, called upon the army to stop supporting despotism and the corrupt political system, and urged workers to 'desist from labour' until their rights as free men were recovered. The lack of soldiers to confront these disturbances and mass marches caused the authorities great worry, and several barracks, such as those in Paisley that were constructed from 1820, were a direct response to these concerns (fig. 6.7).

FIGURE 6.6

In the mid eighteenth century, Piershill House, which stood about a mile east of Calton Hill on the turnpike road leading out of Edinburgh to the east by Jock's Lodge, took its name from its owner, Colonel Piers, who commanded a regiment of horse. It was a large, detached villa with a double coach-house and stabling for eight horses. When it was converted into a barracks in 1793 for regiments of cavalry, the villa became the officers' quarters. The lower ranks and horses were housed together in two newly constructed parallel blocks, facing each other, creating three sides of a quadrangle, with the entrance from the main road to the south. The growing concerns over a possible French invasion gave an official rationale for a trained cavalry corps, even if their value in quelling domestic riots and protesting crowds was perhaps better established.

Over the next century, until its eventual demolition in 1934, Piershill Barracks chiefly entered the news for its insanitary conditions – the Police Committee in 1839 singled it out as 'more unhealthy than any other barracks in the British Isles'. Whether official medical records confirm this seems uncertain, but it was certainly one of the smelliest – sited directly alongside the infamous 'Foul Burn' (the open sewer which serviced most of central and eastern Edinburgh) and, as map (a) shows, just downwind of the open sewage settling ponds at Restalrig. Opinions remained divided on whether the Foul Burn was actually injurious to health, and an official sanitary inspection of the barracks in December 1906 by the celebrated Medical Officer of Health, Dr Henry Littlejohn, made the revealing observation that 'there was nothing that could justify their being classed as uninhabitable on the ground of being insanitary, a great many of the working classes of Edinburgh being very much worse off as regards house accommodation'. In the event, the barracks were declared unfit for cavalrymen – they left for Blair Atholl in 1907, returning to the newly constructed Redford Barracks after 1909 – but, after a modest facelift, Piershill was declared suitable for artillery regiments until its replacement by a housing estate in the later 1930s. Piershill also had its own supporters, and perhaps the final word should go to a nineteenth-century writer, who described 'a stroll from the beautified city to Piershill, when the musical bands of the barracks are striving to drown the soft and carolling melodies of the little songsters on the hedges and trees at the subsession of Arthur's Seat . . . is indescribably delightful'.

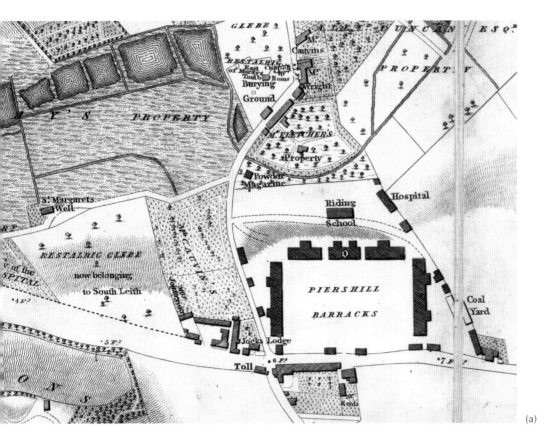

(a)

(b)

Sources: (a) Robert Kirkwood, *Plan of the City of Edinburgh and its Environs* . . . (1817). (b) Ordnance Survey, Large-scale town plan, *Edinburgh, Sheet* 32 (revised 1877).

FIGURE 6.7

(a) The Paisley Infantry Barracks and the accompanying Staff Militia Barracks, on the north and south side respectively of the Glasgow Road in the eastern Williamsburgh district of Paisley, were completed in the early 1820s as part of the response to the Radical War. They are shown clearly on the first edition Ordnance Survey 25-inch to the mile mapping of Renfrewshire, surveyed in 1858. As Paisley was also mapped at 1:500 scale, with 32 large sheets covering the whole built-up area, the internal details of the barracks are also mapped at this scale.

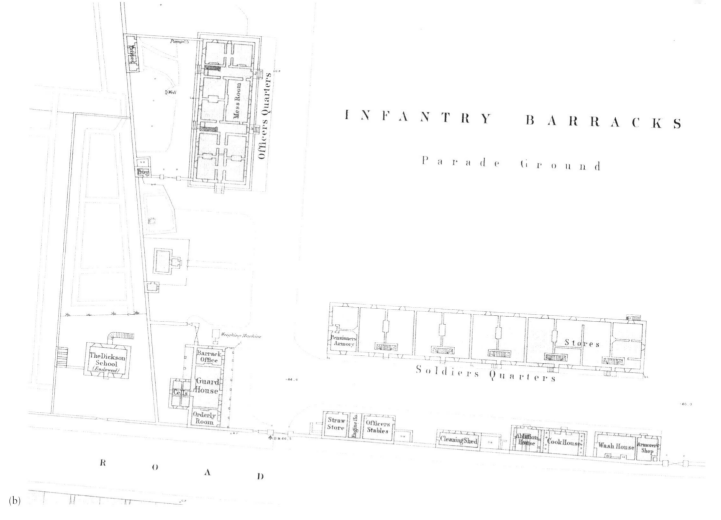

(b)

(c)

(b) For the Infantry Barracks, the officers' quarters and mess room are to the upper left, the soldiers' quarters to the lower right, with accompanying guard houses with cells, offices, stables, ablution house, cook house and armourer's stores.

(c) The Militia Staff Barracks have several of these same functions, including prominent cells to the north, arranged on three sides of a quadrangle. Both barracks were still in use during the First World War, but were subsequently closed and the sites were redeveloped for housing in the later 1930s.

Sources: (a) Ordnance Survey, 25-inch to the mile, *Renfrew, Sheet XII.3* (surveyed 1858, published 1864). (b) and (c) Ordnance Survey, Large-Scale Town Plan, *Paisley, Sheet XII.3.11* (surveyed 1858, published 1864).

Institutional change: the end of the Board of Ordnance and emergence of the War Office

The most important British institutional development for military mapping in this period was the transfer of the Board of Ordnance's mapping responsibilities into the roles of what became Ordnance Survey and the Geographical Section, General Staff (GSGS) of the War Office. Ordnance Survey, originally the Trigonometrical Section of the Board of Ordnance, often traces its foundation to 21 June 1791, when the Duke of Richmond, Master-General of the Ordnance, authorised expenditure on the famous theodolite by Jesse Ramsden. In reality, there were many significant events both before and after this that influenced what would become Ordnance Survey. The Military Survey of Scotland was continually cited by William Roy himself to support the case for a wider British national survey from the 1760s onwards, and had a great influence overseas – for example, on the Murray map of Quebec (1760–61), the Holland survey of the east coast of North America (1764–75), the De Brahm map of Florida (1765–71), the Rennell survey of Bengal (1765–77) and Vallencey's map of Ireland (1778–90). These overseas military surveys strengthened demands for similar work at home, and following France's declaration of war against Britain in 1793 the real concerns over a French invasion added a crucial catalyst. The isolated military surveys of Plymouth and Kent in the 1780s were extended into a broader national survey from 1795. By 1815, most of England south of Birmingham and the English Midlands had been mapped.

Military priorities continued to play a strong influence on Ordnance Survey mapping throughout the nineteenth century, even though their role grew to include civilian needs. This can be seen not only in terms of the aesthetics and style of the maps, reflecting Board of Ordnance conventions, as well as the inclusion of many 'military' features, but also in widespread use of Ordnance Survey mapping by the army and navy. Detailed topographic mapping at consistent scales and based on regular surveys had great value for reconnaissance, training, planning new forts and broader strategic planning of national defences. Administratively, even in 1855 when the Board of Ordnance was formally abolished, Ordnance Survey remained under the control of the War Office, before passing into civil ministerial control under the Office of Works in 1870. The effects of this change can often be overstated. Internally, Ordnance Survey continued with military supervision of a broader civilian staff until 1939, and senior military management of Ordnance Survey formally continued to 1983. There was always close collaboration between Ordnance Survey and the armed forces, and one of the most obvious reflections of this was in the exclusion of secret or sensitive sites. Although some military installations appeared and disappeared from official mapping during the nineteenth century as definitions of 'secret' and 'censorship' changed in response to regular security scares, civilians would have struggled to find their way to Fort George at Ardersier using any pre-1920 Ordnance Survey map (fig. 6.8).

The military reforms from the late 1860s under Edward Cardwell, Secretary for War, which removed the War Office's responsibility for Ordnance Survey, were also responsible for nurturing what would become the major creator of British military mapping in the twentieth century, the Geographical Section, General Staff (GSGS). Originally an Intelligence Department, built around the Topographical and Statistical Department in the War Office (IDWO), it expanded to employ twenty staff in the 1890s, becoming the Topographical Section, General Staff (TSGS) in 1905, and GSGS in 1907. Ordnance Survey and GSGS continued to share information, expertise, mapping and personnel, especially for military mapping purposes. Charles Close, for example, became Chief of TSGS in 1905 before becoming Director-General of Ordnance Survey in 1911. He had previously commanded a survey section in the Boer or South African War in 1901–02, and became Chief Instructor of Surveying at the School of Military Engineering in 1902. In South Africa at this time, as well as in the preparations for and during the First World War, Ordnance Survey undertook substantial drawing and printing tasks for GSGS, and Ordnance Survey base maps were nearly always employed for illustrating Scottish defensive works (fig. 6.9).

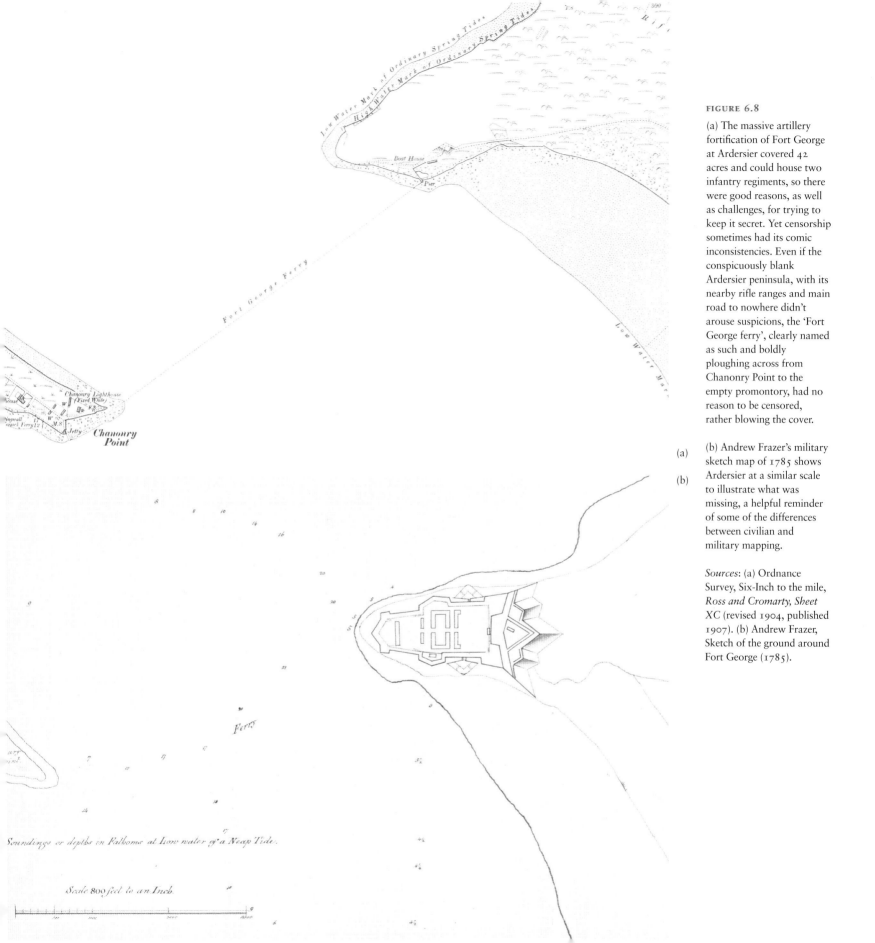

FIGURE 6.8

(a) The massive artillery fortification of Fort George at Ardersier covered 42 acres and could house two infantry regiments, so there were good reasons, as well as challenges, for trying to keep it secret. Yet censorship sometimes had its comic inconsistencies. Even if the conspicuously blank Ardersier peninsula, with its nearby rifle ranges and main road to nowhere didn't arouse suspicions, the 'Fort George ferry', clearly named as such and boldly ploughing across from Chanonry Point to the empty promontory, had no reason to be censored, rather blowing the cover.

(b) Andrew Frazer's military sketch map of 1785 shows Ardersier at a similar scale to illustrate what was missing, a helpful reminder of some of the differences between civilian and military mapping.

Sources: (a) Ordnance Survey, Six-Inch to the mile, *Ross and Cromarty, Sheet XC* (revised 1904, published 1907). (b) Andrew Frazer, Sketch of the ground around Fort George (1785).

FIGURE 6.9

In 1907, the second of three major reports was published on the vulnerability of Edwardian Britain to a seaborne invasion by the Germans, as well as the vulnerability of shipping and the east-coast ports to attack. The military maps that accompanied this report, originally classified as 'Secret', were drawn up at this time by the British War Office, clearly colour-coding the coast into areas of vulnerability to invasion, as well as showing defensive possibilities through hypothetical inland 'fronts', rivers, the communications infrastructure and signal stations.

(a) (*Opposite*) As shown in this detail of the Tay estuary environs, coasts highlighted with thick red stripes were 'practicable for landing', while those with hatched stripes were 'partly practicable', confirming at a glance the major stretches of vulnerable coast south of Arbroath, between Carnoustie and Broughty Castle, and north of St Andrews. These military maps make excellent use of the Ordnance Survey's attractive Quarter-Inch to the mile, 2nd edition colour maps with hachures, giving a useful flavour of relief and terrain, based on a recent revision.

(b) (*Left*) In 1909, more detailed maps using Ordnance Survey, One-Inch to the mile, 3rd edition mapping as a base were drawn up for selected estuaries, including the Tay shown here, highlighting vulnerable parts of the coast for enemy landing and the various defences. As well as showing the Carnoustie Signal Station, they also show the defensive intentions of the electric lights at Broughty Castle, sweeping this narrow, 330-metre wide stretch of the Tay.

Broughty Castle had been fortified in the fifteenth century. After a couple of centuries of neglect and decay, it was acquired by the government in 1855 owing to concerns about Russian warships during the Crimean War. A major reconstruction project in the 1860s, overseen by the young Royal Engineer, Robert Rowand Anderson, was generally considered unsatisfactory for defensive purposes; fortunately for Anderson, he switched to civilian works, for which he was later knighted in 1902. The castle was modified in the 1880s to incorporate the Tay Division Submarine Miners RE (Volunteers), who laid a controlled minefield across the Tay. Three Defence Electric Lights (large searchlights intended to illuminate targets at night) were installed in 1902, and further piecemeal additions were made during both world wars. The maps bear useful comparison with the German Army's mapping of the east coast and its practicability for a seaborne attack in 1940 (see fig. 7.6c).

Sources: (a) War Office / Geographical Section, General Staff, *Map to Accompany the Land Defence of the Scottish Zone* (1907). (b) War Office, *Tay Defences*, revised 1909 (1909).

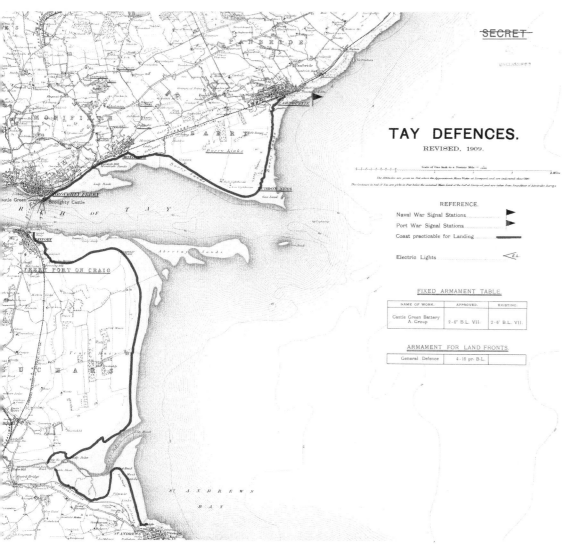

Rosyth naval base

The first of three major reports on the likelihood of a German invasion in Edwardian Britain, published in 1903, led to the Forth being considered a Dockyard Port and Principal Naval Base by a joint Naval and Military Committee. The selection of Rosyth for the naval base was partly based on its excellent defensible situation, with the potential for siting strong batteries further east on the Forth islands and estuary shores. The Admiralty initially purchased 285 acres of foreshore and 1,184 acres of hinterland from the Marquis of Linlithgow in 1903, and construction began in 1909. Rosyth's primary purpose was as a support and heavy repair base for the warships of the Grand Fleet; these included dreadnought battleships which required formidable facilities for their repair and refurbishment. The original scheme comprised a large deep-water basin on the west, entered by a lock with a depth of 36 feet on the sill, two dry docks and provision for a third. Outside was a tidal basin for submarines and smaller craft (fig. 6.10).

The main work was contracted to Easton Gibb & Son Ltd, and was anticipated to take seven years, but there were immense difficulties, particularly in excavating the prevailing heavy clay. Most of the tidal basin and dry docks, as well as the generating and pumping station (to pump water in and out of the dry docks) was complete by 1914. The power station was one of the key buildings in the dockyard, supplying power to the whole complex, including ships, and its industrial chimney (marked as 'conspic.' on all the Admiralty charts) also served as a clock tower. All sorts of ancillary buildings including cranes, workshops and storehouses were required, as well as an oil fuel depot to the east, with a concrete tank storing 250,000 tons of oil, and 37 steel tanks (each with a 5,000 ton capacity). Given the war, there was pressure to bring the dockyard into service; at its peak, a workforce of 6,000 men worked around the clock. The workshops were begun in 1915, the year in which there was an official opening ceremony by the Archbishop of York, but because of a defect in the entrance lock, ships were prevented from entering the main basin until 1916; it was not until 1917 that construction works were finally finished. By this time, the required work in the final year of the war was more limited, although the dockyard witnessed the official surrender of the German High Seas Fleet on 21 November 1918. After it was scuttled in Scapa Flow, many of the ships were salvaged and broken up at Rosyth, which had ideal facilities for this. The war had shown how the torpedo and mine were actually more significant than surface battle fleets, and to an extent, preparations had been made for the wrong sort of war. More seriously still, the post-war years in Britain had non-military preoccupations: short-time working at the yard was introduced in 1921, and the yard closed in 1925, although it reopened again in 1938.

The First World War

The First World War was, more than any previous conflict, a war of maps, with vast quantities produced to illustrate all aspects of the main theatres of war. Scotland was fortunately on the periphery of these main theatres and saw a smaller number of military maps. As with Rosyth, several of these maps were produced for areas planned as new defences. The primary concern at the start of the First World War had been to protect anchorages west of the Forth Bridge, but from 1916, the focus moved to the east. The new lines of defence and batteries on Inchkeith, Inchcolm and Inchmickery took full advantage of the scattering of the Forth islands, and the whole estuary was considered a 'fortress'. Groups of guns and individual batteries on the islands were planned as elements of this wider whole, with interlocking fire, organised under a common control. Inchmickery (a small islet just over a mile north-north-east of Cramond Island) also protected an anti-submarine boom that ran from Burntisland Sands in the north to Cramond Island in the south and was heavily fortified (fig. 6.11).

Although the protection of the Forth occupied a primary place in Scottish military planning and defences at this time, protecting the Clyde, with its important shipbuilding and industrial works, ranked close behind. The Clyde contained

90% of Scottish shipbuilding and marine engineering activity in 1914. Its three main dockyards came under Admiralty control at the outbreak of war, a process that was extended to the rest of the industry under the Munitions of War Act of 1915. Fort Matilda, between Gourock and Greenock, was rearmed with modern guns, a new battery was sited at Portkil, opposite Fort Matilda, on the north shore of the Clyde, and further west a territorial artillery garrison and searchlight was set up at Ardhallow, between Dunoon and Innellan (built 1901–05). Another significant new fortification was the Cloch Point coastal artillery battery, situated three miles south-west of Gourock, on the opposite bank of the estuary from Dunoon, built during 1916–17 by the Royal Engineers and infantry (fig. 6.12). Initially, two guns were removed from Portkil Battery in September 1916 to be mounted at Cloch Point. The Clyde estuary is only about 2.7 km wide here, but the guns had a range of almost 11 km, firing 45 kg shells that could pierce armour. Another function of the battery was to cover an anti-submarine boom strung across the river from Cloch Point Lighthouse.

From February 1915, the Germans began their U-boat campaign in earnest, declaring British waters a war-zone within which merchant shipping could be sunk without warning. This directly contravened the international rules of blockade, with growing numbers of civilian casualties. The sinking of the cruise ship RMS *Lusitania* eleven miles off the southern coast of Ireland on 7 May 1915, with the total loss of 1,200 passengers and crew, caused deep shock internationally. During the course of the war, 89 Scottish boats were sunk while fishing; 15 drifters were sunk off Shetland in a single night in 1915. From 1917, shipping losses accelerated as newer German U-boats were more heavily armed and had a greater range. British marine losses grew to critical levels at this time: in February 2017, 260 boats; March 2017, 338 boats; and in April, a peak of 430 boats were lost (fig. 6.13). The shipping crisis forced the Royal Navy to introduce merchant convoys with escort vessels, which reduced sinkings. The mine barrage between Orkney and Norway also reduced losses in the final months of the war (fig. 6.14).

As the world emerged exhausted at the end of the First World War, many hoped for an enduring interval of peace, but it was not to be. The relatively new technology of the aeroplane, in particular, became a major new offensive weapon, and allied with photography, allowed a dramatically useful new surveillance technology for military intelligence and mapping. The war had also triggered ongoing advances in tanks, anti-tank weapons, machine guns, submarines and radar. Within twenty years, a Second World War brought these combined developments into action, with new geographies of warfare and new types of mapping.

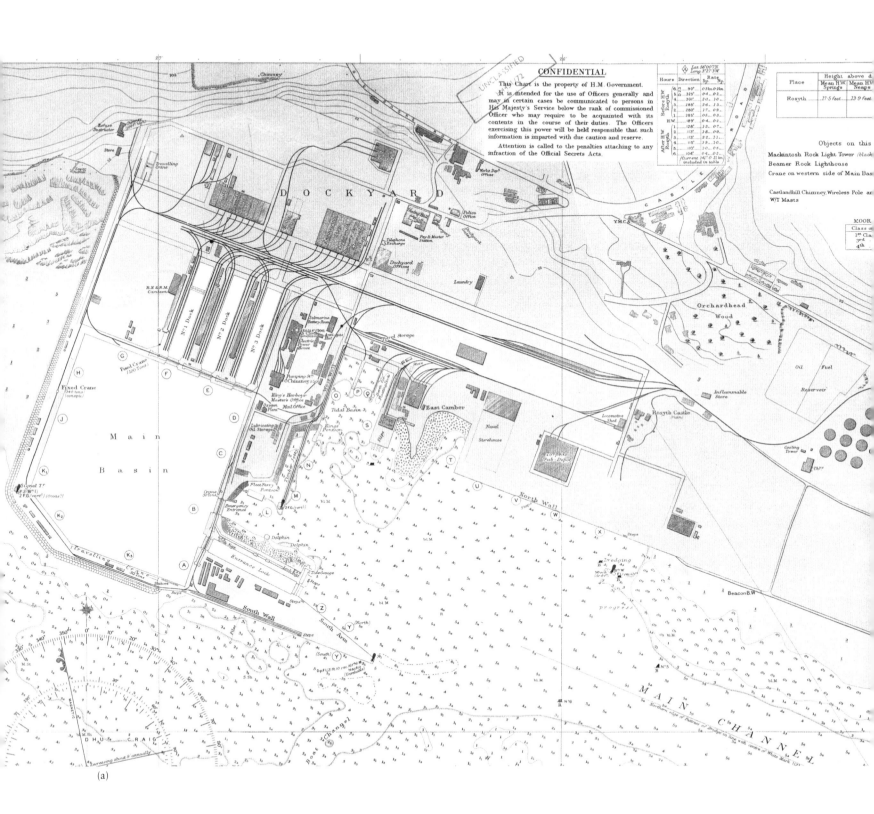

THE FRENCH REVOLUTIONARY WARS TO THE FIRST WORLD WAR

FIGURE 6.10

These three maps contrast the very detailed information on Rosyth available to security-cleared naval personnel only in their Admiralty Fleet charts with the much more limited (or even completely erased) information available to the wider public.

(a) *Admiralty Chart F.86: Approaches to Rosyth Dockyard* is at a natural scale of 1:5,000, allowing all significant buildings to be named, with extensive details – for example, heights of chimneys, berths for ships (shown as letters within circles), and weights of cranes. A 'CONFIDENTIAL' note at the top states:

> This chart . . . is intended for the use of Officers generally and may in certain cases be communicated to persons in His Majesty's Service below the rank of commissioned Officer who may require to be acquainted with its contents in the course of their duties. The Officers exercising this power will be held responsible that such information is imparted with due caution and reserve. Attention is called to the penalties attaching to any infraction of the Official Secrets Acts.

This chart has edits and minor corrections to 1948, but substantially reflects Rosyth soon after its completion in 1917.

(b) The far less detailed *Port Edgar to Carron: Admiralty Chart 114c* reflects what was available to non-naval personnel at the same time. It is at a natural scale of 1:18,260, 3.6 times smaller than the Fleet chart, and only names a few of the significant buildings by comparison. It also gives a more dated impression of progress, and completely omits the oil depot.

Both maps (a) and (b) were based on the same surveys by Captain J. W. F. Coombe in 1916–19 and Commander C. H. Knowles in 1921–23, but the latter chart had substantial generalisation and a reduction of content.

(c) The Ordnance Survey Six-Inch to the mile map, revised in 1924–25, manages to omit even more, just giving a fringe of detail to a large blank space. Curiously, the large rectangular buildings which escaped the censor's knife were the Torpedo Sub-Depot and the Naval Storehouse; on the north side, Castle Road curves down to a police station before disappearing into the void.

Sources: (a) Hydrographic Office, *Admiralty Chart F.86: Approaches to Rosyth Dockyard* (1944). (b) Hydrographic Office, *Port Edgar to Carron: Admiralty Chart 114c* (1925). (c) Ordnance Survey, Six-Inch to the mile, *Fifeshire XLIII.NW* (revised 1924–25, published 1928).

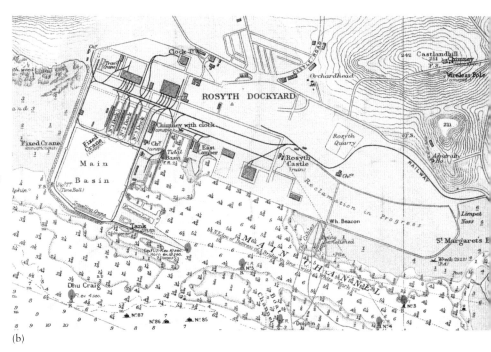
(b)

(c)

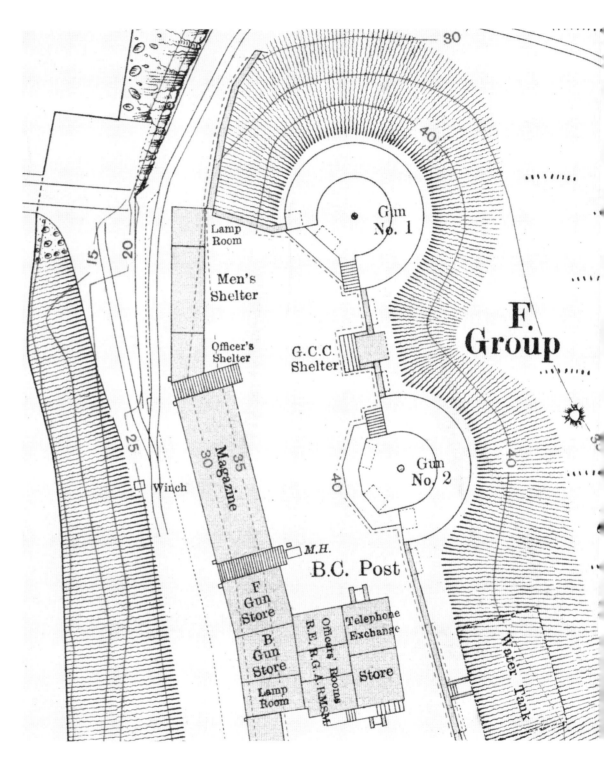

FIGURE 6.11

(*Left*) This very detailed plan of Inchmickery by the War Office at 1:360 or 30 feet to an inch, with red contour lines every 5 feet, was surveyed in 1914, with revisions in 1918, and was intended for classified military use only. It allows the terrain and defences to be scrutinised in minute detail, with seemingly every inch of the island remodelled for military purposes. Inchmickery was first manned in March 1915, when a detachment of 72 non-commissioned officers (NCOs) and men landed on the two islands from Leith Royal Garrison Artillery. The large concrete gun emplacements shown here originally housed four 12-pounder guns on the highest ridge of the island, but these were upgraded with guns transferred from the island of Inchgarvie, below the Forth Rail Bridge, in late 1916. Behind the guns were large magazines, stores and shelters for officers and men (see detail from main map), as well as latrines and a sewage pipe. Initially there were two searchlights but these expanded to four 'Electric Lamp Emplacements' by the end of the war: one in the south, two on the upper east side and one in the north. Although the battery garrison was not entirely self-sufficient, it had been planned as a practical, working battery with an engine room and cooling tanks, messes, canteens, stores, accommodation for officers and sergeants, a plumber's shop, carpenter's shop, smithy, ablution rooms, bath rooms and cook houses. Some of the guns were removed in 1917 and the rest in 1924, but the island was refortified, with several defences rebuilt, in 1939–42.

Source: War Office / Geographical Section, General Staff, *Plan of Special Survey Inchmickery, Firth of Forth* (1918).

NOT TO BE PUBLISHED   CLYDE. CLOCH POINT SHEET 1.

FIRTH OF CLYDE

E.L. Emplacement No. 3

Guard Room

Bar. No. 2

R.G.A. Headquarters Office

Bar. No. 3

Cloch Lighthouse

E.L. Emplacement No. 4

Cloch Point Battery

A Group

B.C. Post & E.L.D.

FIGURE 6.12

This very large-scale plan of Cloch Point, 5 miles west of Greenock, by the War Office was surveyed in 1904, with revisions in 1918, and bears close comparison with fig. 6.11. They are both at 1:360 or 30 feet to an inch, with red contour lines every 5 feet, allowing a very detailed view of the military infrastructure. Both were declassified only in the 1980s.

(a) (*Left*) This map shows the westernmost of two sheets which together show the whole extent of the site, this sheet showing the battery to the lower centre, and to the north, part of the temporary accommodation for the garrison in wooden huts, with a cook house and ablution rooms. The Cloch Lighthouse, dating from 1797, with its foghorn and engine room, is shown to the west of the battery, halfway down the left-hand edge of the sheet.

(b) (*Right*) This shows the detail of the battery itself, with its two guns at either end of a continuous row of buildings offering basic shelter for officers and men, stores and latrines. The battery remained in use between the wars; more permanent quarters for troops and officers were built in 1939–40.

Source: War Office / Geographical Section, General Staff, *Clyde, Cloch Point Sheet 1* (1918).

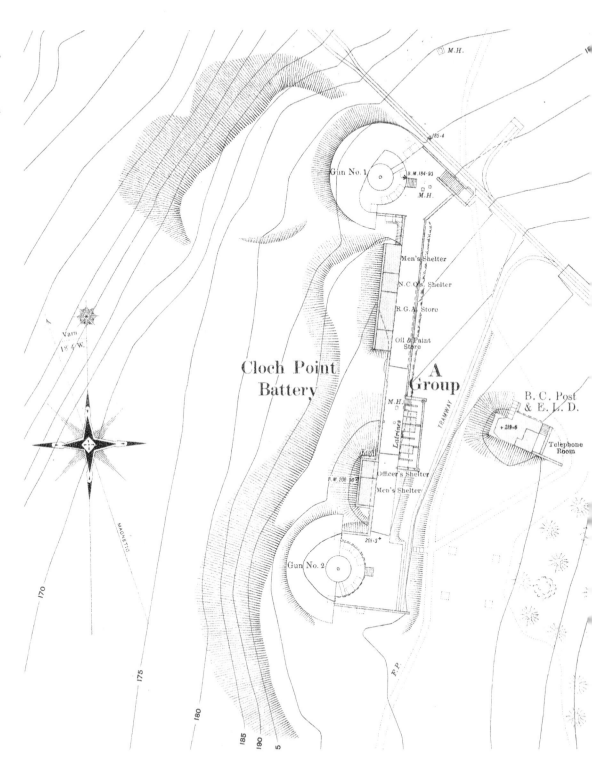

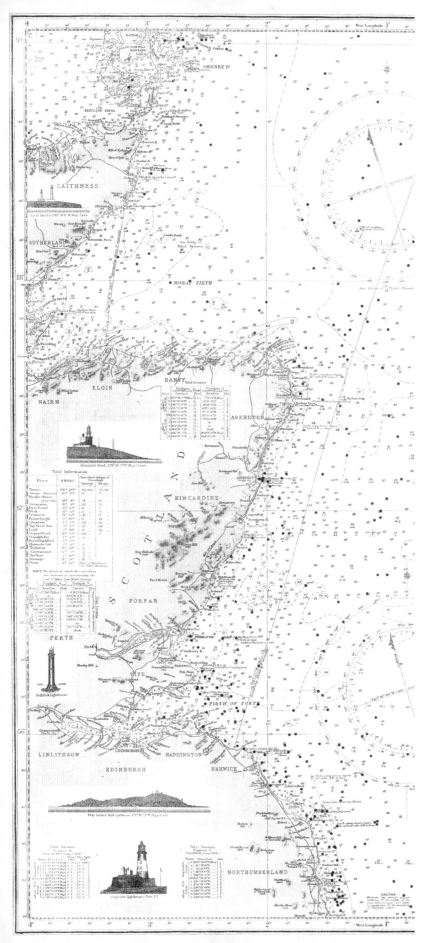

FIGURE 6.13

(*Left*) This Hydrographic Office wreck chart is one of seven which collectively cover all the British home waters, attempting to plot the positions of wrecks at sea (shown as red dots), which were a continual marine hazard in the years after the First World War. Although the east coast of Scotland is not as heavily peppered in red as the southern North Sea or English Channel, the scale of the losses is clearly visible.

Source: Hydrographic Office, *British Isles, English Channel and North Sea. Wreck Chart in 7 sheets. Sheet II. D20* (1919).

FIGURE 6.14

(*Right*) During the First World War, Germany laid more than 40,000 mines, which claimed around 500 merchant vessels, as well as 44 Royal Navy warships and 225 naval auxiliary vessels. Much of the German North Sea coast, with its naval bases at Cushaven and Wilhelmshaven, was heavily defended by minefields (Zone IV on the map). The response around Scottish waters initially involved laying more localised minefields to defend key ports and estuaries, and it was only in 1917 that a large minefield was laid from Orkney to Norway (Zone II). A similar mine barrage had already been placed across the English Channel, which had resulted in U-boats diverting north around Scotland. The main objective was therefore to prevent U-boats from reaching the North Atlantic and attacking trans-Atlantic shipping, bringing supplies to Britain. The North Sea Mine Barrage or Northern Barrage was largely constructed by the United States Navy (assisted by the Royal Navy) and proved a costly, technically challenging and dangerous exercise. It no doubt hindered U-boat movements in the closing months of the war, and official statistics credited it with the certain destruction of four U-boats, and probable destruction of four more.

This chart shows the danger areas in 1919, when equally challenging and dangerous work was underway to try to remove the mines. From April to November 1919, Royal Navy minesweeping efforts involved 600 officers and 15,000 men on 421 vessels, around one third of which were damaged by exploding mines. The main darker patches 'hatched and tinted' red were known to contain minefields, while those lighter tinted red areas were 'considered dangerous owing to the possible existence of moored mines'. Rear Admiral Strauss of the US Navy, who commanded its minesweeping operations, declared the North Sea barrage free of mines by 30 September; he was subsequently made a Knight Commander of St Michael and St George for his efforts. In October 1919, however, twenty crewmen drowned when the Swedish steamship *Hollander* sank after striking a mine, while on 1 December 1919, the steamer *Kerwood* struck a mine and sank. Many doubts about the effectiveness of the minesweeping work persisted through the twentieth century.

Source: Hydrographic Office, *Waters Surrounding the British Islands: Mined Areas and Safe Channels* (1919).

FIGURE 6.15

(*Overleaf*) War Office / Geographical Section, General Staff, Map to Accompany the Land Defence of the Scottish Zone (1907).

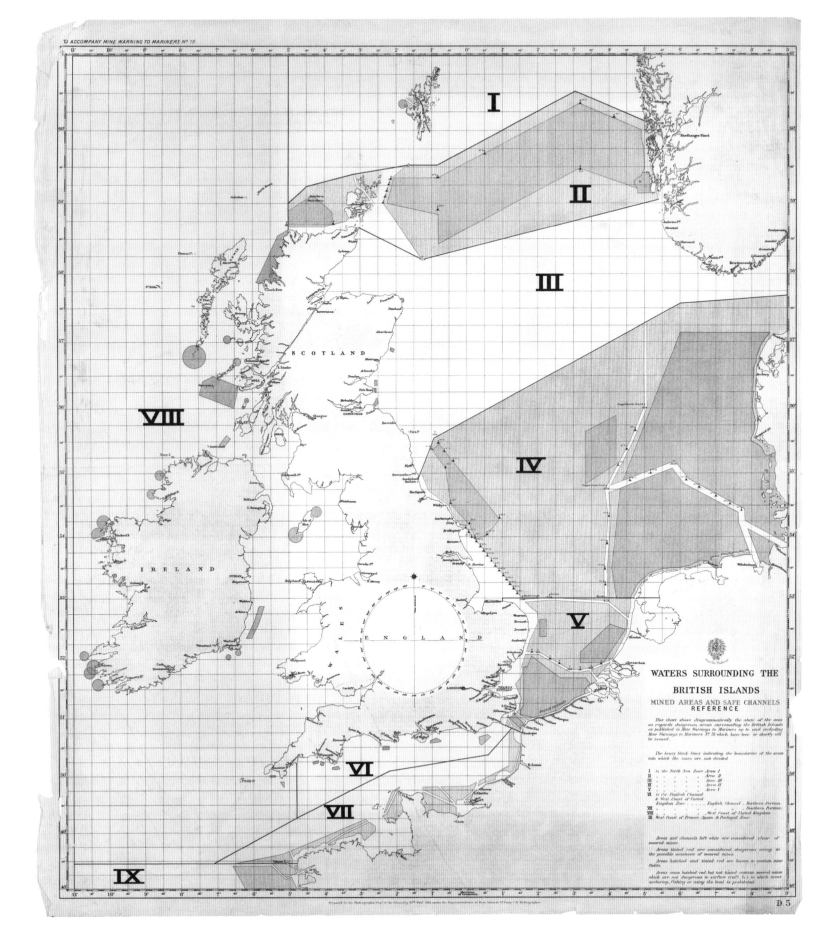

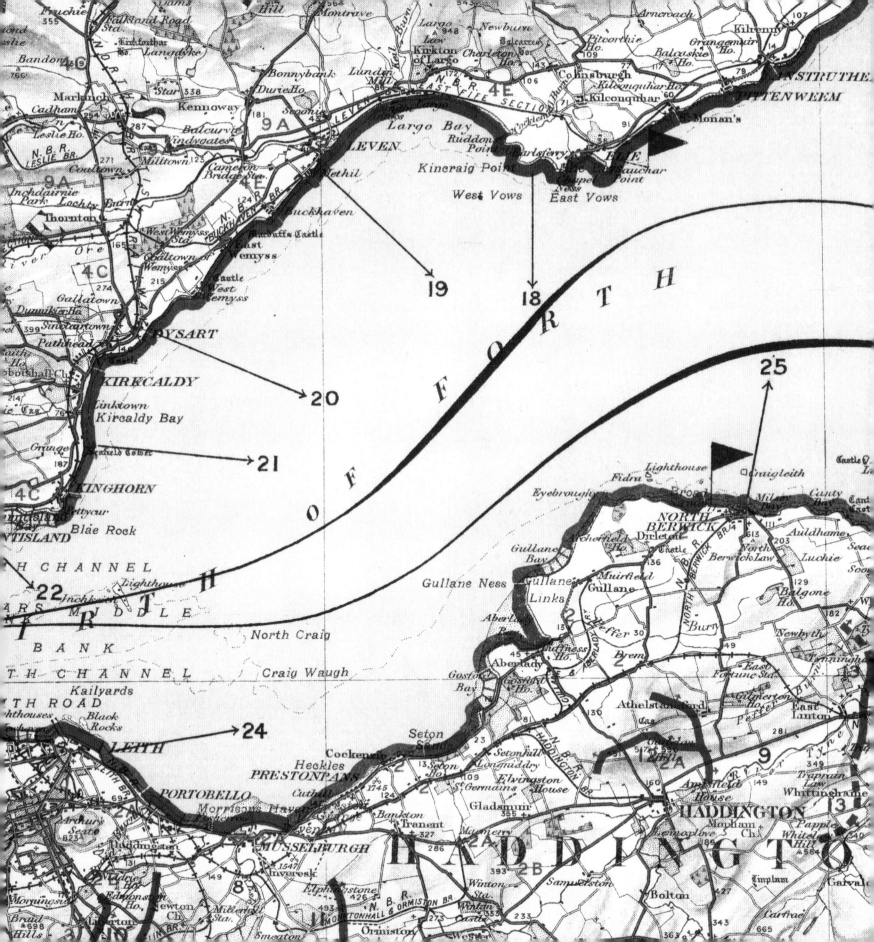

CHAPTER SEVEN
THE SECOND WORLD WAR TO THE PRESENT DAY, 1939–2018

The period since the start of the Second World War has seen the most profound changes in military technology, in the geography of warfare and in mapping. In terms of weaponry, the steady rise of the range and speed of the aeroplane and submarine throughout the twentieth century, along with the later growth of nuclear weaponry and long-range missiles, completely altered the practicalities and geographies of war. In Scotland, the earlier concerns about naval shipping, invasion beaches and coastal defences up to and during the Second World War were rendered largely irrelevant with the advent of nuclear missiles from the 1950s. These changes took place against a changing background of perceived enemy threats, from Germany and its allies during the Second World War, to the Soviet Union during the Cold War from the 1950s.

During the twentieth century, map-making was assisted by a mass of new information which became available through aerial reconnaissance, supplemented by satellites projected into ever higher orbits through rockets. There was a growing overlap between photography and mapping, and in the last half-century, satellite information came to serve as the basis for new weaponry through GPS (global positioning systems) locational information and computer-controlled firing and targetry. This period has also witnessed huge volumes of mapping initially on paper, and then in digital form. During the Second World War, Ordnance Survey produced around 300 million maps for the Allies, while the American Army Map Service produced more than 500 million maps, just two institutions among many others producing maps at this time. The rise of digital mapping technologies and geographic information systems from the 1970s was particularly prompted by military needs. Today, while the paper topographic map is very much still in evidence for military purposes, a huge value is placed on geographic intelligence that is monitored, processed and acted on by machines.

This chapter looks first at some of the British military maps in the build-up to the Second World War, before examining German mapping for attack and invasion of Scotland in the Second World War. A briefer section looks at post-war mapping, particularly for aviation, as well as Russian military mapping in the 1980s.

Preparations for war

Between the world wars, the British army struggled for funding when much of the British population fervently hoped that nothing was worth the price of another war. There were more cuts to the army than to the air force and navy, as bombardment from the air and the threat of foreign submarines to maritime commerce were seen as the principal new threats to national security. Against this background, Ordnance Survey and the War Office struggled in the face of massive cuts to resources and personnel, and lack of coordination between the services hindered matters further. The result was that the topographic base for British mapping became steadily less up to date, while research into new techniques, such as photogrammetry from aircraft, lagged behind that of other European countries. Despite persistent lobbying in the 1930s by Ordnance Survey's Director-General Malcolm Macleod for a suitable military map of the United Kingdom at 1:25,000 scale, at the outbreak of war in September 1939, this series only covered one seventh of the country. As an emergency measure, its production was dramatically speeded up, at a rate of fifty maps a day, so that the whole series of around 2,000 maps could be produced in about two months. It was a classic rush-job, with no time for surveying or updating (fig. 7.1).

During the First World War, the development of grid systems on mapping, allowing the artillery to fire with precision on remote targets, had proved its effectiveness, not least in battles such as Cambrai in November 1917. Before acting as Director-General of Ordnance Survey from 1935, Macleod was head of the War Office mapping authority (or Geographical Section, General Staff) from 1929, and he continually stressed the importance of mapping for artillery purposes, having seen its effectiveness first-hand in the First World War. He strongly believed in the value of direct survey for mapping, with suitable projections and appropriate geodesy, and this had an important impact on the development of British military mapping. His ideas were partly supported by Colonel John Fuller – one of the foremost experts on military engineering in Britain in the interwar period – who defined the 'ideal military map'. This included all the common geographical features (roads, railways, rivers, villages, mountains and forests) with the addition of a host of strategic, tactical and military features for new mechanised warfare. Fuller suggested, for example, that the ground should be coloured to show where tanks and other off-road vehicles could move at ease, so that a military commander would know at a glance where to deploy mechanised arms. Other new features of importance were existing and potential aerodromes, prevailing winds at various times of the year (in case gas attacks were to be used), petrol supply and petrol stations, repair centres, motor works and machine shops. In a mechanised war, these would be as important as wells, water holes and oases in desert warfare. We see these features appear on some British twentieth-century military maps, and they were very much in evidence in German wartime mapping and Russian post-war military mapping.

Scotland's topography and geography lent itself well to the training of service personnel throughout the twentieth century. From a British military perspective, the rugged, mountainous Highlands with challenging weather conditions, surrounded by a complex coastline and islands, with a variety of sea conditions and strong tides, were enhanced by Scotland's relatively sparse resident population and the relative inaccessibility of the area for foreign continental enemies to see. Around a quarter of a million Allied soldiers received special training in Scotland during the war, many in the Arisaig and Morar Protected Area (out of bounds for anyone without the relevant passes), supplemented by many other sites, including Derry Lodge in the Cairngorms, Arran, Inveraray and Largs. Much of the training for the D-Day amphibious landings took place on Scottish beaches. There were also dedicated military training camps, such as Barry Buddon Ness – around 2,300 acres (930 hectares) of dunes, foreshore and rough ground with isolated stands of trees, jutting out into the Tay between Monifieth and Carnoustie. It was used from the mid nineteenth century by the Forfarshire Rifle Volunteers, by the Panmure Battery of the Forfarshire Artillery Brigade, and by a Royal Naval Reserve Battery. In 1897, the land was sold by Lord Panmure to the War Office, and it has remained in its ownership ever since, still serving today as a major training centre for infantry and light artillery, with around 30,000 soldiers passing through the camp every year (fig. 7.2).

Primarily owing to censorship, the steady increase in the number of airfields during the twentieth century is difficult to track on maps, but evidence can be found (fig. 7.3). During the First World War, more than 500 airfields were created on

FIGURE 7.1

(a) (*Right*) This War Office map, known by its military series number of 'GSGS 3906', covered the whole country at 1:25,000, deemed to be an essential scale in the event of the anticipated German invasion. The maps were produced by reducing topographic information from the latest available Ordnance Survey Six-Inch to the mile maps down to 1:25,000 (the thin white edge of one of these sheet lines can be seen running horizontally across this extract). A rectangular military grid was added, and second editions of the maps (as shown here) photo-enlarged contours from the One-Inch to the mile maps (often not quite matching the Six-Inch contours) and overprinted them, often in brown. The maps' distinctive qualities reflect severe wartime constraints of time and resources.

These constraints were also helpful in omitting to show new munitions factories and related military infrastructure. This particular map shows the site of the Hillington Rolls-Royce factory (which lies to the east of the 'Scottish Industrial Estate'), opened in 1937 with support from the Air Ministry. The first Merlin engines were being produced two weeks before the start of the Second World War and, by 1943, the factory was producing nearly 400 engines a week. It is reckoned that the factory produced over 23,000 aircraft engines during the war. Although a few non-military details were updated on these maps, the requirement to produce the maps fast, without revision, played a useful censorship role – the latest Ordnance Survey Six-Inch to the mile maps revised in 1938–39 omitted the Rolls-Royce factory as well as the active RAF Renfrew Airfield, dating from the First World War.

(b) (*Opposite*) The 'cartographic silences' on the 1941 map are revealed by this uncensored Ordnance Survey topographic mapping after the War. The later mapping, appropriately generalised for this scale, is also much easier to read than the photographically reduced GSGS 3906.

Sources: (a) War Office / Geographical Section, General Staff, Great Britain, 1:25,000 GSGS.3906 2nd Provisional edn, *Sheet 29/68 S.W.* (1941).
(b) Ordnance Survey, 1:25,000 Provisional edition, *Sheet NS56* (revised 1938–55, published 1958).

(a)

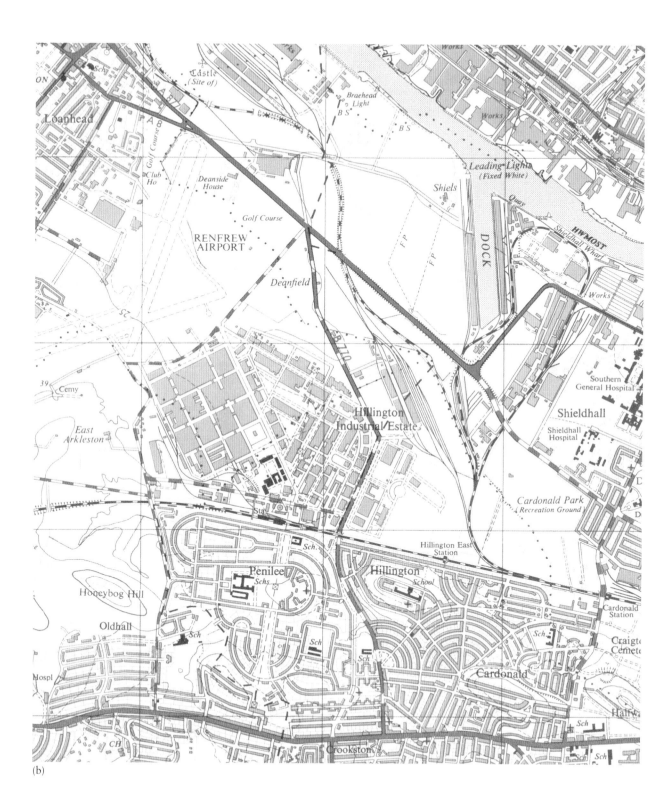
(b)

(a)

(b)

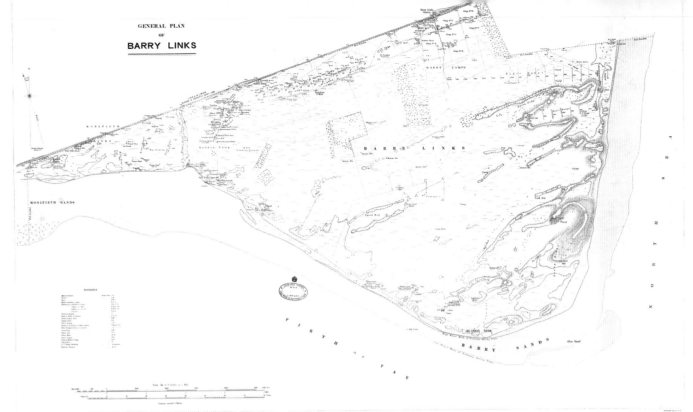

FIGURE 7.2

(a) This plan of Barry Buddon Ness by the War Office dates from 1939 and is based partly on the Ordnance Survey Six-Inch to the mile topographical mapping of 1923, but with updates and enlarged to Nine-Inch to the mile (1:7,040). Apart from the Dundee to Arbroath Railway to the north, and the Barry Buddon Ness Lighthouses in the south, virtually all other man-made features shown have a military purpose. The long range of numbered camps in the north, running alongside the railway, had few permanent buildings, but extensive demarcated locations for cook houses, wash houses and segregated latrines for officers and men. The main rifle ranges were in the east, the longest stretching 1,700 yards, with artillery ranges and huts, bomb-proof shelters and stores all lying to the south.

(b) The main set of permanent buildings of the camp itself were in the centre west, including the officers' quarters and mess, many storage buildings, stables, a hospital and hutments, and an RAF landing ground. Several rifle ranges and target sheds lay further to the west.

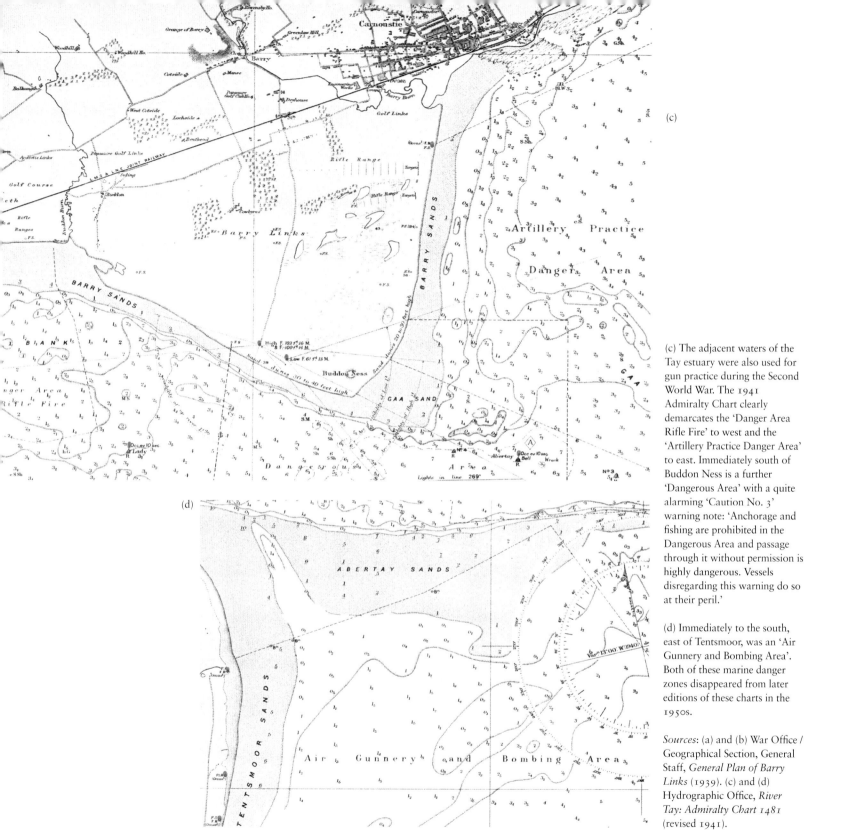

(c) The adjacent waters of the Tay estuary were also used for gun practice during the Second World War. The 1941 Admiralty Chart clearly demarcates the 'Danger Area Rifle Fire' to west and the 'Artillery Practice Danger Area' to east. Immediately south of Buddon Ness is a further 'Dangerous Area' with a quite alarming 'Caution No. 3' warning note: 'Anchorage and fishing are prohibited in the Dangerous Area and passage through it without permission is highly dangerous. Vessels disregarding this warning do so at their peril.'

(d) Immediately to the south, east of Tentsmoor, was an 'Air Gunnery and Bombing Area'. Both of these marine danger zones disappeared from later editions of these charts in the 1950s.

Sources: (a) and (b) War Office / Geographical Section, General Staff, *General Plan of Barry Links* (1939). (c) and (d) Hydrographic Office, *River Tay: Admiralty Chart 1481* (revised 1941).

(a)

FIGURE 7.3

Donibristle Aerodrome was developed from September 1917 primarily to maintain both land-planes and seaplanes for the navy. It extended over 53 hectares by 1918, with several large aeroplane sheds, ammunition stores and accommodation buildings. It was linked to the shore by a railway line, two miles long, which ran out onto a pier for seaplanes, to transport them to and from the aerodrome. Donibristle was one of only two Scottish aerodromes that continued in active military use after the First World War through to the Second World War, closing in 1959.

Because of censorship, finding detailed evidence of the aerodrome on publicly available maps is difficult, as for most Scottish military airfields.

(a) Donibristle was named, along with only two other military airfields (Leuchars and Turnhouse), on the first Ordnance Survey 'Popular' edition maps in the 1920s. Despite a tightening of national security in 1924 (which saw various forts, dockyards, and other military sites erased from maps), the Ordnance Survey successfully resisted military pressures to have aerodromes removed, cleverly invoking the potential for civil aviation as an argument for on-going depiction. This resulted in a brief 'golden age' (1925–34) when the great majority of active aerodromes were shown.

(b) Commercial map-makers like Bartholomew, keen to keep their maps up to date, specifically showed these new airfields with a prominent red aircraft symbol.

(c) Thereafter, security restrictions with the looming threat of war led to erasures, and there is no trace of Donibristle airfield in the Ordnance Survey Popular edition mapping (with National Grid) in 1945, nor on any other Bartholomew mapping. By 1950, it is estimated that across Great Britain, of the 1,375 sites created during aviation's major growth phase, and well embedded in the landscape, only 240 (17%) made any discernible appearance on Ordnance Survey mapping.

Sources: (a) Ordnance Survey, Scotland, One-Inch to the mile, Popular edition, *Sheet 68, Firth of Forth* (revised 1923–26, published 1928). (b) John Bartholomew & Son, Half-Inch to the mile, *Scotland, Sheet 8 Forth* (1934). (c) Ordnance Survey, Scotland, One-Inch to the mile, Popular edition with National Grid, *Sheet 68, Firth of Forth* (1924 with later revisions, 1945).

(b)

(c)

British soil, but two-thirds of these were fairly humble landing grounds or seaplane slipways with no major landscape legacy. In the run-up to the Second World War, there was a further dramatic expansion and the creation of dedicated aeronautical charts (fig. 7.4a). The RAF's preferred series of charts for flying had evolved from the 1920s, originally using the Ordnance Survey's Quarter-Inch to the mile base mapping, steadily stripping the underlying map of unnecessary detail, and overprinting them with a growing range of additional details for aviation in red (fig. 7.4b). By 1939, this overprint included thirty different symbols that a map sheet might include, compared with just six used in 1925–26.

During both the First and the Second World Wars, there was a dramatic expansion in the publication of the basic black-and-white newspaper map, illustrating immediate events of the war at home and far away. The major growth in the circulation and sale of newspapers in the First World War, and the need to visualise the geography of events as they unfolded, demanded the use of maps to an unprecedented extent. By the end of the First World War, most newspapers either had their own in-house cartographers, or maintained close links with commercial cartographers, ensuring that the growing prominence of newspaper mapping continued. While the vast majority of these maps depicted overseas locations, there were also maps published of Scotland – for example, planning evacuation from the large cities (fig. 7.5).

The scale and speed of the proposed emergency evacuation from Scottish cities at the outset of the Second World War was quite astonishing, with proposals to evacuate nearly 1.8 million people. It was assumed Glasgow could be evacuated in just three days, Edinburgh and Dundee in two days, and official guidance implied detailed planning:

> Train and bus time-tables have been worked out in consultation with the Traffic Commissioners and the railway companies, and Local Authorities in reception areas have been notified of the train arrival times and of the maximum number of persons likely to be sent to their areas.

The reality of evacuation was, of course, quite different from that implied by the neat, ordered lines of the evacuation map. Perhaps only a tenth of the original estimate, around 180,000 people, left their homes, and two-thirds of these returned by March 1940 when the anticipated mass-bombings had failed to materialise. The practicalities on the ground were often chaotic, and profoundly dislocating for everyone involved. The evacuation forced people and families together who were often completely divided on class, social, sectarian, urban/rural and ideological lines. That said, it highlighted many problems of structural inequality, and was an important influence on post-war attitudes to housing and society.

German invasion and bombing maps

Following the successful German invasions of Poland (September 1939), Norway (April–June 1940) and France (May–June 1940), Hitler's attentions turned to Britain, particularly from July 1940. On 16 July 1940, Hitler issued Führer Directive No. 16 for preparations for a landing operation against Britain, known as 'Operation Sealion'. Significant resources were devoted to the project, particularly in research and documentation, with maps, photographs and detailed written reconnaissance.

The German army's publication *Militärgeographische Angaben über England: Ostküste – (Nördlicher Teil vom Humber bis zum Firth of Tay) / Military-geographical Information on England: East Coast – (Northern Part from the Humber to the Firth of Tay)* gives a good flavour of the high quality of this information. As the Germans lacked comprehensive intelligence from spies, their information was primarily drawn from British civilian topographic maps and photographs, supplemented by limited aerial reconnaissance. The bulk of the publication consists of photographs, supported by maps, which collectively give an excellent impression of the landscape and terrain that the army might find itself in. The photographs were often of key military targets (fig. 7.6a and b), with their locations shown on the accompanying maps which described the nature of the coastline for an invasion from the sea (fig. 7.6c and d). An accompanying land cover map was specially drawn, showing the characteristics of the terrain for moving tanks and other armoured vehicles around (fig. 7.6e). This volume formed part of a larger series covering European countries, which was generally available only to generals at the highest level and their immediate staff.

So the German army was excellently briefed on the military geography of Britain and on coastal defences. Although the further they might have advanced inland, the more sparse their knowledge became, it was still comparable to that of the British army. As part of its preparations for potential invasion or occupation, the German army reprinted standard topographic mapping by Ordnance Survey. Ironically, the map includes a few features, including the military aerodrome at Leuchars, which were subsequently erased from later Ordnance Survey mapping from the 1930s onwards for censorship reasons, so the German use of this earlier mapping, obtained before wartime restrictions, was more useful for military purposes (fig. 7.7).

British Second World War mapping

On the nights of 13 and 14 March 1941, 236 German bombers dropped 1,700 incendiary bombs and 270 tons of highly explosive munitions on the Clydebank environs. It was a carefully planned operation, using excellent maps and reconnaissance photographs taken before the war, supported by a new and effective radio navigation system, and assisted by the clear moonlit night. The aircraft arrived in three waves from shortly after 9 pm to 3 am. Initially, pathfinder aircraft from north-west France dropped incendiary bombs to create an inferno and light up the targets. Heavy bombers then followed in subsequent waves from France, the Netherlands and Norway, using approaches from the south by Liverpool and then the north over Loch Lomond to disguise their main objective. The incendiary bombs set fire to distilleries, the Singer timber yard, and the large Admiralty oil storage tanks at Dalnottar, which burned for days. Parachute bombs were used too, exploding at roof level with a much more destructive impact. In the chaos and carnage down on the ground, the town was evacuated as much as was possible because the authorities rightly expected a second night of raids.

By the evening of Saturday, 15 March, probably over 40,000 people (more than two-thirds of the population) had left the town. The Department of Health was on the scene soon afterwards. Over the following months, it produced detailed situation reports with maps showing bomb damage (fig. 7.8). By the end of April, it was calculated that 55,000 people had been made homeless in Clydebank and the greater Glasgow area. In total, only seven houses out of a stock of 12,000 remained undamaged, with around 4,000 destroyed, 4,500 severely damaged and 3,500 in the serious-to-minor-damage category. Official figures (possibly an underestimate), record that the Clydebank raids killed 528 people and seriously injured 617, with totals of around 1,200 people killed, and 1,100 seriously injured in the whole of Clydeside.

The Germans' main targets in Clydebank were the munitions factory in the former Singer sewing machine works, the shipyard of John Brown & Company and Beardmore's engine works. In practice, although Singer lost a valuable stock of timber, and the engine works of Aitchison Blair were completely destroyed, other works suffered only partial damage. Following repairs over the following months, it was reported that production returned to near-normal levels, with the main restraint being the labour force, who were understandably reluctant to return.

Less than two months later, there were further Luftwaffe raids on Greenock on 5–7 May, with the main targets being the Royal Naval Torpedo Factory and RAF Greenock, a seaplane maintenance base. Greenock was also an assembly point for the Atlantic convoys. The damage was far less than in Clydebank owing to better civil defences, a fighter squadron from RAF Ayr which shot down some enemy aircraft and a 'starfish decoy' near Loch Thom that tricked the bombers.

During the war, the Clyde estuary was full of British and American shipping, not just for the Atlantic convoys, but also

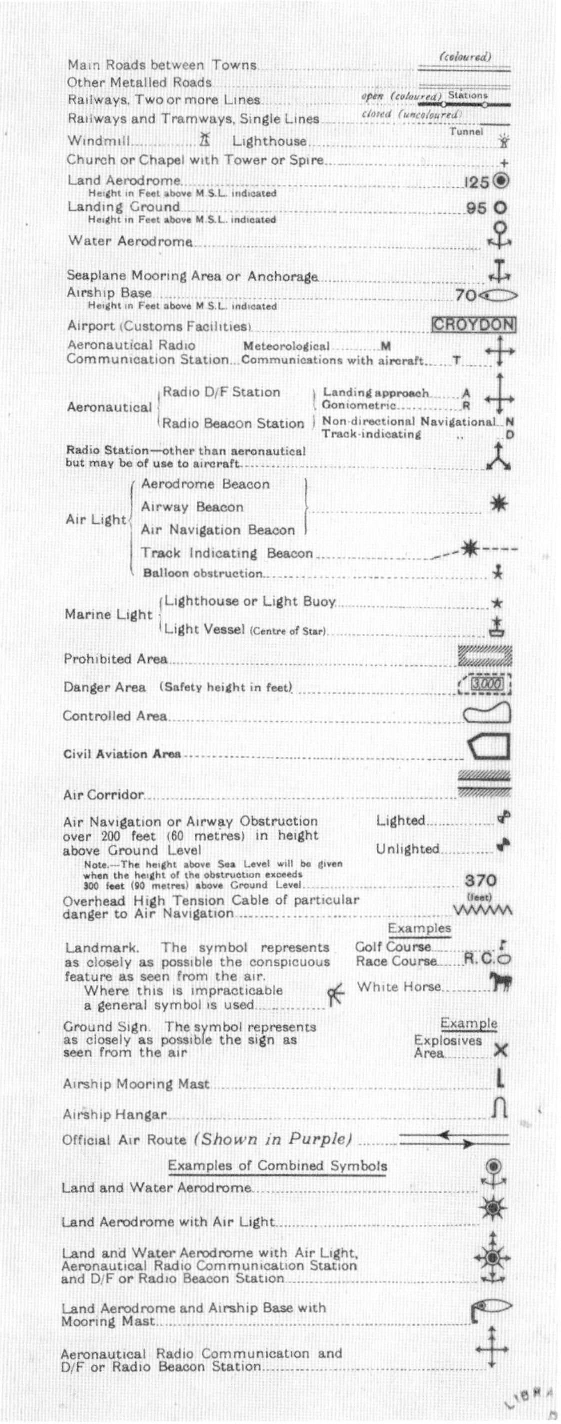

FIGURE 7.4

This detail of the Forth from a map printed at the outbreak of the Second World War gives a good example of most detailed aviation maps that the Royal Air Force preferred to use at Quarter-Inch to the mile scale (1:253,440), with overprinted aviation information in red. On 16 October 1939, the RAF squadrons at Drem and Turnhouse shot down the first German aircraft on mainland Britain, and this is the kind of map they may have had to help them. Perhaps the most obvious overprinted information shown here were the extensive 'Artillery Ranges' in the Forth, with a red dashed 'Danger Area' around them. The number in the red box shows that this rose to 5,500 feet, along with a more localised 'Bombing Range' south-east of Donibristle, dangerous up to a much higher level of 30,000 feet. There was also a 'Mortar Range' at the top of the Pentlands around Allermuir Hill, dangerous up to 3,000 feet. Aerodromes or landing strips were also necessarily obvious, with formal aerodromes, such as Donibristle, Turnhouse or Drem (drawn in on this map), shown by a dot within a circle; more limited landing grounds, such as Tranent, were shown by a circle only. Note, too, the 'Seaplane Mooring Area or Anchorage' east of North Queensferry with its anchor symbol. The Luftwaffe christened the Firth of Forth a 'suicide alley', partly due to the comprehensive artillery defences shown here and fighter aircraft nearby. Stars dotted around the coast indicated lighthouses, numbered inside the map cover with an accompanying key describing their specific flashing sequences. Other features that were very visible from the air, such as golf courses, show how extensive these had become by this time, while the 'dumb-bell' symbol indicated masts or 'Air Obstructions' over 200 feet high. Visually, one of the most striking changes to the series, introduced in 1935, was to show the layer-coloured relief in purple (rather than shades of brown as used previously), which was much easier to see in the subdued amber cockpit lights which helped airmen to keep better night vision. These maps were folded for easier use in the cockpit, using the 'Michelin fold': each sheet was folded once down a horizontal centre line and then concertinaed, allowing successive areas to be read without unfolding more than necessary.

Source: Ordnance Survey, Scotland, 1:253,440, Royal Air Force Edition, *Sheet 3 – The Forth & Tay*. (1939).

FIGURE 7.5

This map was published in *The Scotsman* on 5 September 1939, two days after the formal declaration of war by Britain against Germany. It displays evacuation plans which had been developed over the previous year by a commission chaired by Sir John Anderson, MP for the Scottish Universities. There had been trial attempts to evacuate people during the summer of 1939, but these were largely unsuccessful, and it was only the formal declaration of war that prompted people into action.

The map zoned Scotland into three main types of area: sending areas, neutral areas and receiving areas. The main urban conurbations of Glasgow, Clydebank, Edinburgh, Rosyth and Dundee, shown by a dot within a circle, were the sending areas, requiring evacuation. Nearby these centres were neutral areas, shown in black – relatively densely populated already – which would neither send nor receive people. The receiving areas were of course the rural areas, allocated roughly to each conurbation as adjacent county clusters, so that they were accessible by rail within a day. Owing to the much larger population of Glasgow and Clydebank, their receiving areas occupied a wide swathe of counties, from Wigtownshire in the south-west to the Buchan coast in the north-east. The counties north of the Great Glen were considered as 'reserve areas', which would not receive people, owing to a combination of the long travel distances to reach them, as well as the relatively limited accommodation available.

Source: 'Evacuation map of Scotland', *The Scotsman*, 5 August 1939, p. 16.

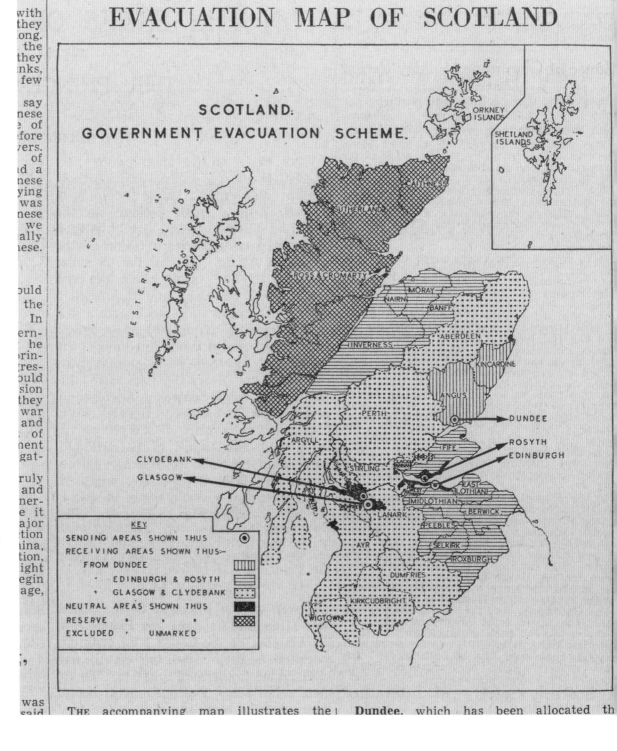

FIGURE 7.6

In preparation for 'Operation Sealion', the invasion of Britain, detailed topographic intelligence was hastily compiled by the German army in 1940, and produced as a book entitled *Militärgeographische Angaben über England: Ostküste* (*Military-geographical Information on England: East Coast*), including air photos and maps:

(a) An air photo of the Forth Bridge, a key infrastructure target for German bombers, giving its height above the water, and also details of the headroom and distance between the two arches at high tide.

(b) An air photo of Bo'ness, correctly referring to it as 'Coal export port, 7.5 nautical miles above the Forth Bridge on the south bank of the Forth (10,000 inhabitants)'.

(a)

102. Blick von Queensferry über die Forthbrücke (113 m Höhe) zum Nordufer des Firth of Forth.
Die Durchfahrtshöhe der beiden Brückenbogen über Hochwasser beträgt auf 146 m breiter Strecke 45 m.

(b)

103. Borrowstouness (Bo'ness).
Kohlenausfuhrhafen, 75 sm oberhalb der Forth-Brücke am Südufer des Forth. (10 000 Einw.)

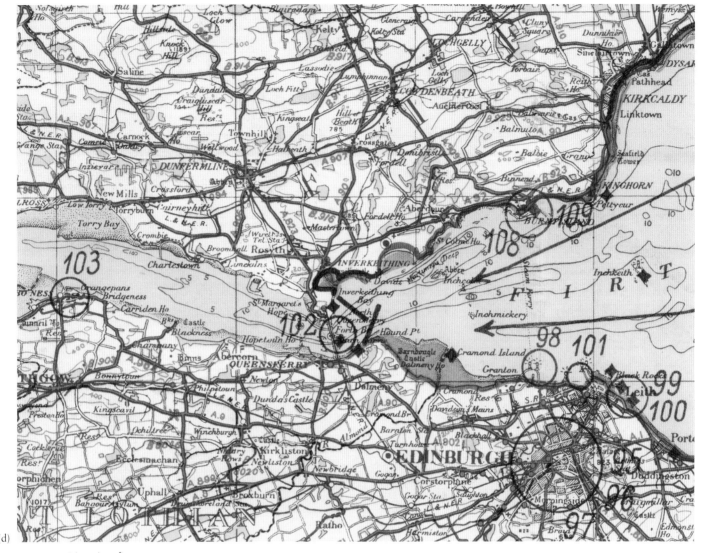

(d)

FIGURE 7.6 (*continued*)

(c) (*Left*) The locations of the air photos (figs. 7.6a and b) are clearly shown as purple circles with numbers on the accompanying Ordnance Survey Quarter-Inch to the mile maps, with the coastline colour-coded to show the type of coast for landing. The red overprint around the coastline indicated cliffs and steep gradients, while green and yellow (e.g. south of Carnoustie) indicated flat and sandy shores respectively.

(d) (*Above*) This more detailed section from the overprinted Ordnance Survey Quarter-Inch to the mile maps (c) correctly picks out with purple diamonds all the Firth of Forth coastal batteries and defences (in place at the beginning of July 1940).

(e) (*Overleaf*) The book also contained a useful map showing land cover or terrain, translated as 'Soil characteristics in the hinterland of the east coast between the Humber estuary and Firth of Tay (northern part)'. This map classified land characteristics, showing its suitability for different types of army vehicles, aircraft landings and building materials, with the main lines of roads and railways shown with black continuous and dashed lines. Most of the eastern part of Scotland was clay (orange: Lehm), interspersed with igneous/crystalline rocks (red: Kristalline Gesteine) and sandstone (dark green: Sandstein). The invasions of Norway, Belgium and France had already proved how vital this ground information was, particularly for tanks, and at a glance the map allows an army commander to assess this.

Sources: (a) 'Blick von Queensferry über die Forthbrücke . . . zum Nordufer des Firth of Forth'; (b) 'Borrowstouness (Bo'ness). Kohlenausfuhrhafen, 75 sm oberhalb der Forth-Brücke am Suduferdes Forth (10 000 Einw.)'; (c) and (d) Ordnance Survey, Quarter-Inch to the mile, 4th edition map, overprinted with landing beach categorisation and locations of photos; (e) 'Bodenverhältnisse im Hinterland der Ostküste zwischen Humbermündung und Firth of Tay (Nördlicher Teil)' all from Generalstab des Heeres, *Militärgeographische Angaben über England: Ostküste – (Nördlicher Teil vom Humber bis zum Firth of Tay)* (1940).

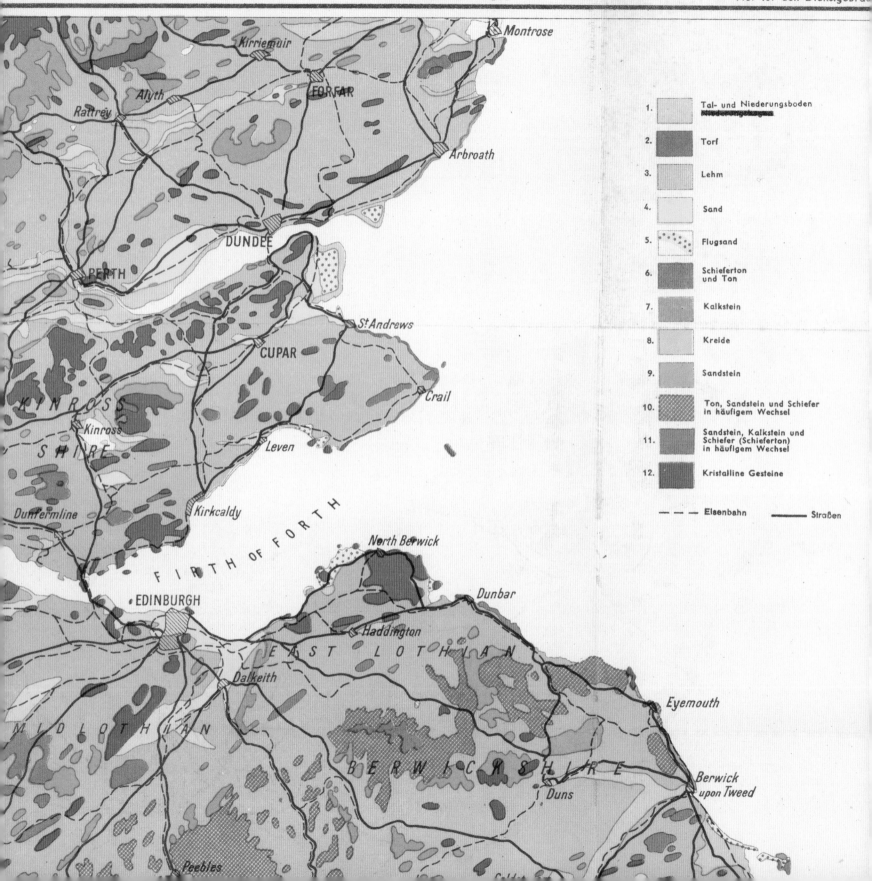

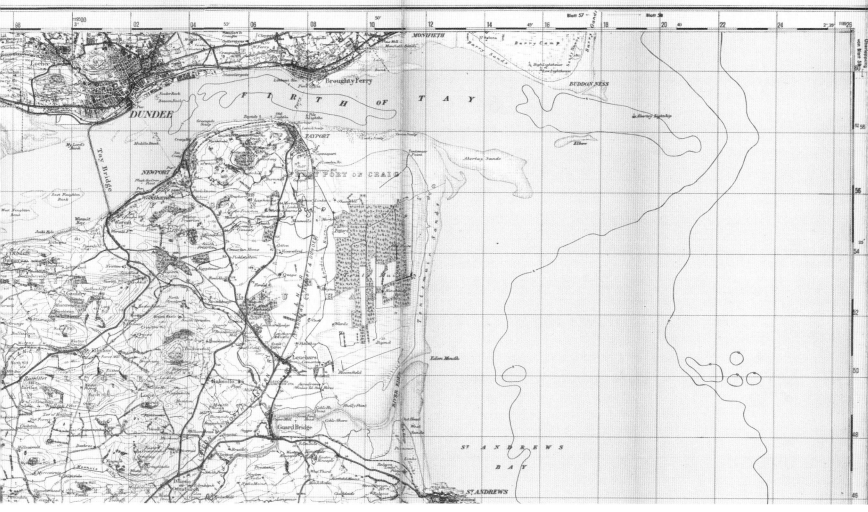

FIGURE 7.7

This is a detail from *Blatt Nr 64 – Dundee und St Andrews*, part of the German Army's 1:50,000 series, which metricated and enlarged Ordnance Survey's One-Inch to the mile (1:63,360) 'Popular edition' mapping of 1928. This work was a wartime 'rush job', with no time or intention to update topographic information. Apart from enlarging the scale, the only other alterations were the translation of marginal information and the map's legend. The note above the legend provides guidance on converting contours and heights, which German users would have expected to see in metres: 'All height information, including contours are given in English feet. To roughly convert to metres apply: 3 feet = 1 metre. All height information on the map must therefore be divided by 3 in order to obtain metre information. For more detailed conversions see table.'

Source: Oberkommando des Heeres, *Schottland 1:50,000, Blatt Nr 64 – Dundee und St Andrews* (1941).

FIGURE 7.8

This detail is from one of thirty-five maps covering all of Clydebank, which were compiled by the Department of Health in the weeks after the Clydebank blitz. Given the official censorship of mapping showing the devastated area (which would have been highly useful to the Germans, as well as useful propaganda for anti-military groups), these are a rare example of detailed maps showing the exact positions of bombs across the town. They confirm that civilian housing bore the brunt of the bombing, while industrial premises in fact escaped relatively lightly. The street outline and buildings were traced from Ordnance Survey 25 inch to the mile maps, with the position of bombs dropped on 13–15 March and 5–7 May taken from the Burgh Surveyor Records.

Source: 'Department of Health, Burgh of Clydebank. Air Raids – 13/15th March – 5/7th May 1941. Bomb Nos: 132 to 158. Position of Bombs taken from Burgh Surveyor Records', from *Record of High Explosive and Incendiary Bombs dropped on Scotland during the War* (1941). Courtesy of the National Records of Scotland.

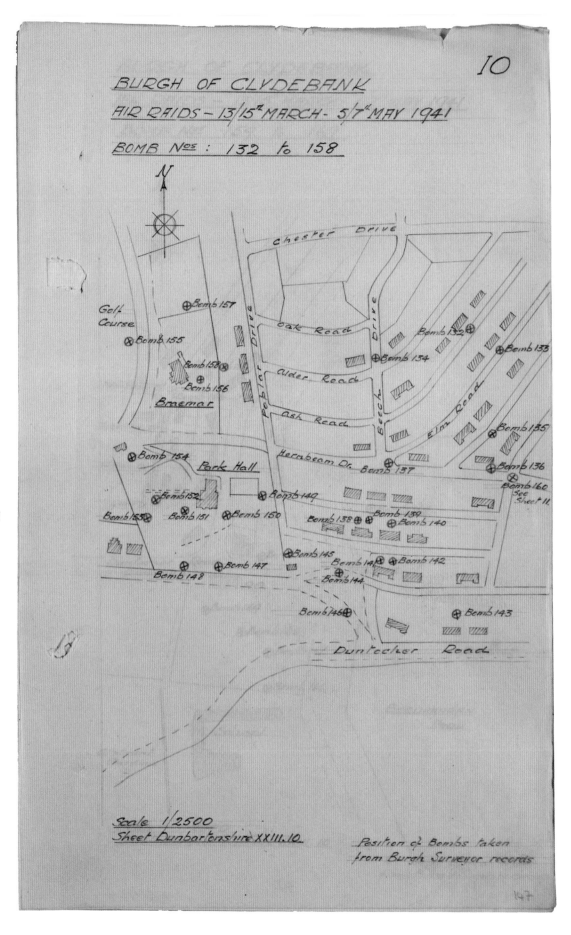

for training and repair purposes. From March 1941, Rosneath Bay on the Gareloch was chosen as a suitable site for a new United States naval base, with its excellent deep-water anchorages nearby, and the onshore spacious grounds of Rosneath estate (fig. 7.9). Originally the intention was that it should be used for US destroyers and submarines if and when America entered the war, but after Pearl Harbor in December 1941, most of the US destroyers and submarines were sent to the Pacific, so Rosneath became available to British forces from April 1942. The base was lavishly constructed to American standards, and used intensively for training purposes, including the American amphibious forces in preparation for the proposed Allied invasion of north-west Africa, code-named 'Operation Torch'. The base was also used for similar training in the build-up to D-Day in 1944.

Air photo mosaics and aeronautical charts

Towards the end of the Second World War, it became clear that widespread detailed mapping of the United Kingdom was required to plan post-war reconstruction, even though this was a major task that would take many years to complete on the ground. As an interim measure, and to make use of some of the highly skilled RAF fighter pilots seeking work, an extensive aerial photographic survey of Great Britain was initiated from 1944, called 'Operation Revue'. For selected urban areas, Ordnance Survey created and published special 'air photo mosaics', so-called because of the rectification of the original air photos and careful blending of overlapping sections to form a composite layer. In theory, and perhaps initially, this was a more honest exercise of representing the landscape below as it was – but within a few years, there appeared a number of interesting 'revised editions' (fig. 7.10).

In the post-war period, aeronautical charts were increasingly sophisticated, tailored towards all aspects of military aviation: flight planning, cruising, descent, low-level flying, approaches to airfields and landing. Scotland became a major zone for military aviation, with extensive low-level flying ranges, bombing ranges and RAF bases. In contrast to the initial development of aeronautical charts which developed after the First World War using standard topographic maps as their base and featuring an overprint of aeronautical information (fig. 7.3), by the 1960s there were many more bespoke maps for aviators (fig. 7.11). On the one hand these newer maps reflect the considerable simplification of topographic detail, so that only very selected features in the landscape, of value for air navigation, are shown. On the other, they also reflect the major efforts to standardise symbols, colour and text on aeronautical charts internationally. A major part of the remit of the International Civil Aviation Organisation (ICAO), founded in 1947, was the standardisation of scales, symbols and formats for aeronautical charts. In addition, military mapping organisations in NATO countries (including the British Army of the Rhine, who published this map) comply with NATO Standardization Agreements and the Air Standardization Coordinating Committee (ASCC) Air Standards, partly based on ICAO standards, too.

Russian military mapping

Russian military maps have a distinguished pedigree, and in the twentieth century they became the most extensive, detailed military mapping ever produced, with global coverage. Military concerns were integral to their development. The Russian Military Topographic Directorate (VTU) was founded in 1812, five months before Napoleon's disastrous invasion of Russia. Then in the wake of the Bolshevik Revolution in 1917, a massive expansion in state cartography took place, given further impetus by the Second World War. By 1957, the entire territory of the Soviet Union had been mapped at 1:100,000 (over 13,000 sheets), while a much larger project to map the whole world was well underway. It is only following the disintegration of the Soviet Union in the 1990s that many of these formerly secret maps of parts of Britain and Scotland became generally available in Western, non-military circles, allowing something of their history and content to be studied.

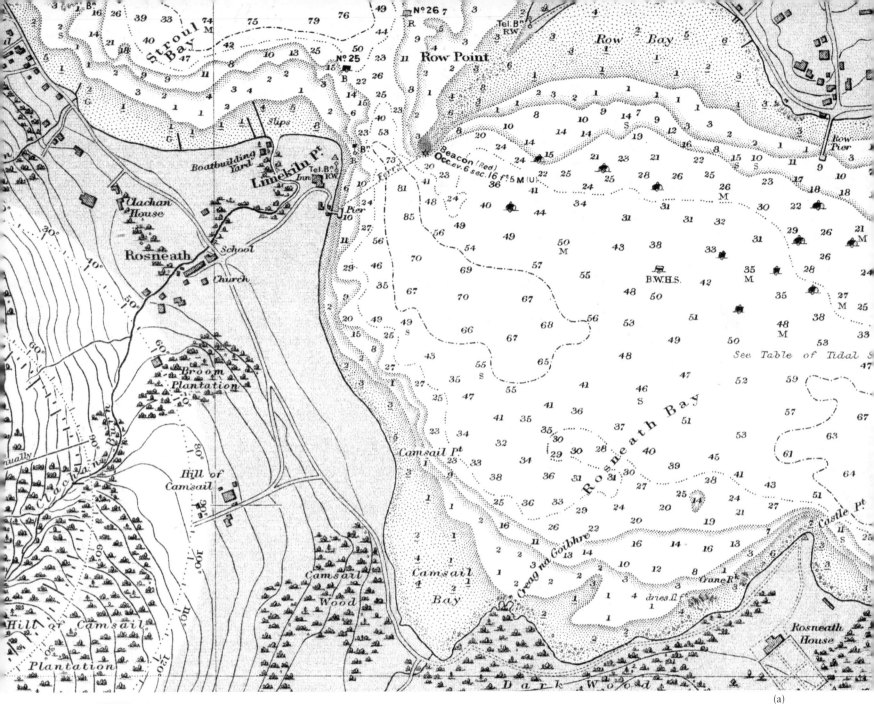

(a)

FIGURE 7.9

These two Admiralty charts of the Rhu narrows at Gareloch, the first dating from 1922 and the second from 1943, show the transformation of the area with the siting of Rosneath naval base from 1942. United States civilians collaborated with British Royal Engineer corps in the major construction work involved. Rosneath House to the south became the American staff headquarters and the planning centre for 'Operation Torch' as thousands of US navy, marine corps and army personnel arrived at the base to practise amphibious landing techniques at Rosneath and around the Clyde sea lochs. In February 1944, over 6,300 officers and men were stationed here. In 1946, a proposal to renovate the base facilities for use as a combined operations establishment was considered but then cancelled, and in 1948 the base was closed.

Sources: (a) Hydrographic Office, *Gareloch: Admiralty Chart 2000* (1922).
(b) Hydrographic Office, *Gareloch: Admiralty Chart 2000* (1943).

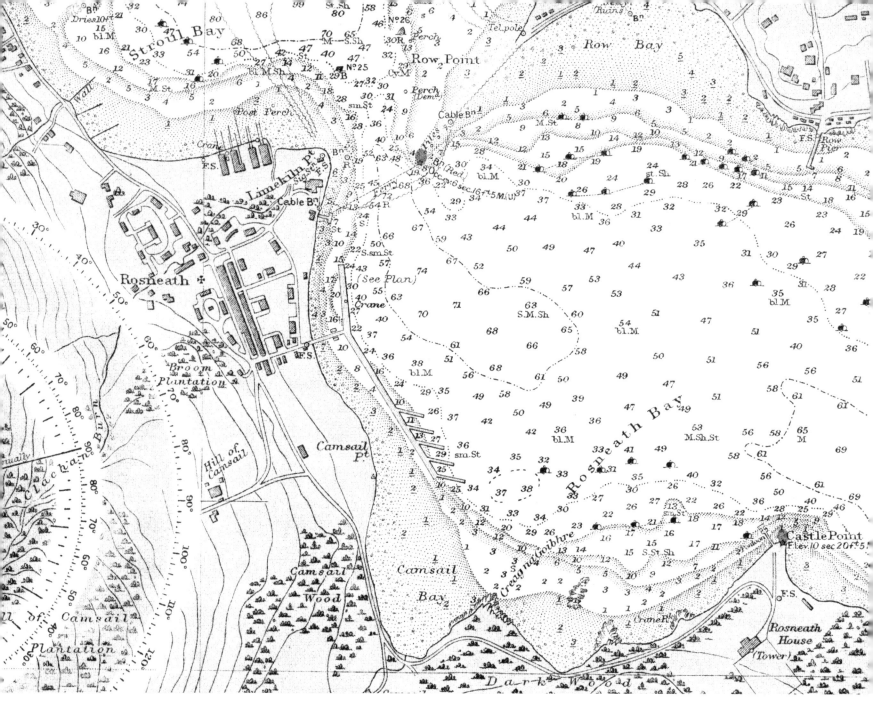

(a)

FIGURE 7.10

These two 'air photo mosaics' show changing attitudes to censorship in the post-war world, as well as the doctoring of air photos. The mosaics were originally intended for official use only, but were offered for sale to the public between 1945 and 1947 in what turned out to be an unsuccessful effort to recoup costs.

(a) The first photo shows good detail of a military installation in an uncensored manner – the Anti-Aircraft Battery at Limekilnburn, around three miles south of Hamilton on the A723 Strathaven road. The battery formed part of the Clyde Anti-Aircraft Defences, and included a 'Gun Laying' radar platform (the central rectangular structure shown here), surrounded by four large gun emplacements. A command post and other buildings can also be seen, along with huts which would have formed the main accommodation camp, to the north-east of the battery. During the 1940s, there were major advances made in radar sensors, and the Mark II radar, introduced here, could allow bearing measurements as accurate as half a degree, twice as accurate as the Mark I radar, along with better elevation measurements, and just within the range needed to correctly aim the guns.

(b)

(b) By the late 1940s, however, there were growing concerns that some mosaics showed military installations all too clearly, and many, including the one in fig. 7.10(a), were withdrawn from public sale. From 1950, newly doctored mosaics were re-issued for these selected locations, with a false landscape of empty fields and suitably erased military infrastructure. Security concerns grew and, in March 1951, libraries were warned that the original 'true' mosaics should be withdrawn from public use. In 1954 the mosaics were withdrawn from sale completely. Limekilnburn's gun platforms were reused after the Second World War in the 1950s as a Cold War Battery, and although the site still doesn't appear on standard Ordnance Survey maps, the structures are clearly visible on satellite imagery today.

Sources: (a) Ordnance Survey, *Air Photo Mosaic 26/75 S.W.* (1946).
(b) Ordnance Survey, *Air Photo Mosaic 26/75 S.W.* (1950).

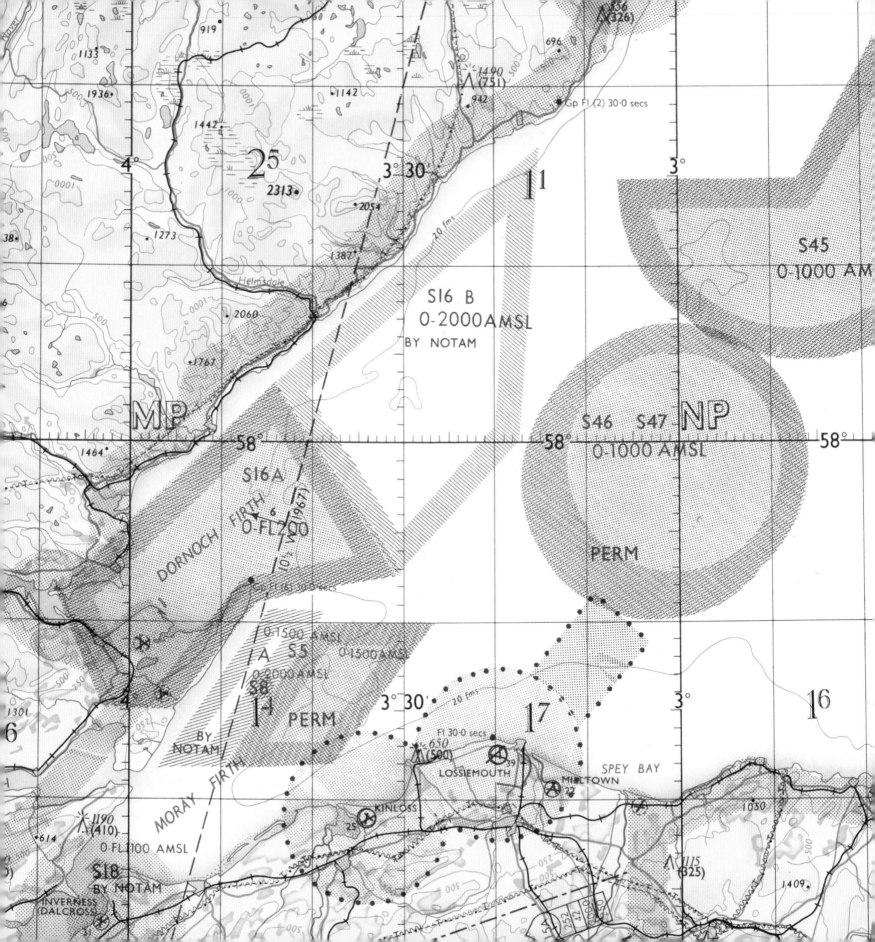

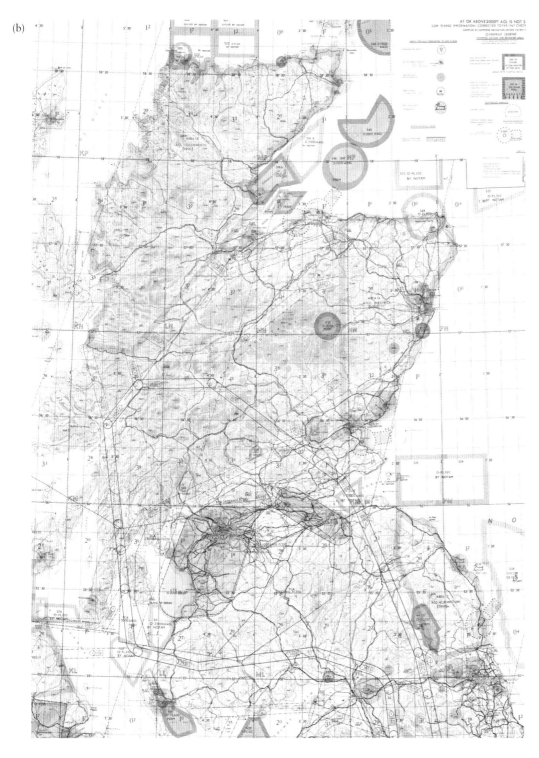

FIGURE 7.11

(a) This sheet forms part of a European military aeronautical series from the 1960s and is an excellent example of the type of specialised air chart that developed after the Second World War. It also shows well the extensive restricted areas for military low-flying, as well as the large number of military airfields in and around the Moray Firth at this time.

The prominent red-tinted areas were all restricted danger areas to low-flying aircraft; those with central red tints were permanently active. Each zone contained its own alphanumeric Area Designation (e.g. S16, S47), the lower and upper limits of the area (e.g. 0–2000 AMSL, meaning up to 2,000 feet above mean sea level) and the times of activity, with PERM meaning permanent. The authorities for these areas were also sometimes given, many citing the *Notices to Airmen* (NOTAM) service, which also issued updated information for updating charts.

Active airfields were shown as blue circles (the circle representing a 6,000-foot radius) with the runway pattern also indicated within the circle. The number beside each airfield gave its height in feet above sea level. Blue-hatched areas with perimeter dots showed the Military Airfield Traffic Zone around the major military airfields of RAF Kinloss and RAF Lossiemouth.

The topographic detail of the map is deliberately limited. Bold blue dots at prominent locations along the coasts represent lighthouses, with their individual flashing sequences (Fl) given. Red zigzag lines are electricity power lines. Other ground-level obstacles are triangular red mast symbols, with heights in feet recorded – the first number is the height of the top of the mast, while the second number (in brackets) is the height of the mast itself. In addition to the shaded relief, there are spot heights for principal elevations, and a very prominent regular graticule of degrees. The diagonal blue dashed lines show the isogonal lines of magnetic north.

(b) The whole map sheet shows the uneven distribution of these low-flying zones across Scotland, with other clusters to the north and east of Cape Wrath, in south-west Scotland, especially Luce Bay and off Dundrennan, with another zone in the outer Forth estuary.

Source: British Army of the Rhine, *Ascent Topographic Low Flying Chart: Europe 1:500,000. Sheet 7, United Kingdom, North* (1967).

Soviet military maps cover the world at a set of scales from 1:1 million to 1:10,000, and at each of these scales, they conform to a consistent specification, using the same standard symbols, projection and grid. The immense value of this for facilitating their interpretation and use by military personnel cannot be overemphasised; military commanders or soldiers in wartime scenarios had no need to cope with a foreign cartographic map series, with its own alien style, symbols and language. This rational, consistent way of controlling and managing the world cartographically used lines of latitude and longitude as sheet lines, with sheet names following the International Map of the World. The basic 1:1 million scale quadrangle (e.g. O-30, including Scotland (fig. 7.12a, b)) spans an area of 4° latitude by 6° longitude. O-30 lies in band O (fourteenth band north, 56° to 60° north) and in zone 30 (0° longitude to 6° west). The 1:500,000 scale maps (e.g. O-30-Б (fig. 7.12c)) are based on subdivisions of this sheet into four, covering an area of 2° latitude by 3° longitude, using the first four letters of the Russian alphabet, labelled А, Б, В and Г. Larger scales are based on further subdivisions, with 1:100,000 maps based on 144 subdivisions. City plans such as Aberdeen carry the name of the city and their parent 1:100,000 sheet (e.g. O-30-104 (fig. 7.12d)).

The city plans are also accompanied by a Справка (*spravka*) – a detailed geographical description of the city. For a city such as Aberdeen, this is some 1,780 words, covering the physical geography, geology, soils and relief, climatic conditions, transportation and communications, water supply, sewerage and utilities, economic base and chief industries, medical institutions, notable buildings and overall layout. It makes fascinating reading, not only because of how detailed, informative and accurate it is, but also for the particular features recorded that are of clear military value, and it gives useful supporting information on the purposes behind the mapping. Under the 'City Precincts' section, for example, we read that:

The landscape of Aberdeen suburban area represents partly open undulating coastal flatlands, with patchy hills (50–150 m. altitude), dissected by deep river valleys that are the major obstacles for non-road mobile machinery. Mud season lasts from October to April; during this period transport mobility is hampered.

The attention to detail on the road network is also revealing:

A dense network of motorways in the precincts of Aberdeen ensures vehicle traffic in all directions throughout the year. Highways are improved with bituminous concrete or asphalt surface; width of the carriage way is 7–9 m., the main body of the road is up to 20 m. wide. The rest of the roads have asphalt or black macadam pavement and are up to 11 m. wide. Stabilized soil roads (up to 6 m. wide) are graded; the roadbed is reinforced with gravel and macadam. The bridges are predominantly stone, steel-concrete or metal.

The nature of the coastline, in terms of cliffs, dunes and depth of water, was also accurately described, along with the note 'the coastal area north from Aberdeen is suitable for amphibious landing'. It also correctly notes 'Numerous stone quarries, mainly granite, in Aberdeen suburbs can be used as shelters. The largest stone quarry is Rubislaw.'

Finally, and perhaps not surprisingly, Aberdeen harbour is a particular focus of the 'Industrial and Transport Facilities' section:

Aberdeen seaport (located in the estuary of Dee river) is the major maintenance base for oil deposits in the North Sea. A 6 m. deep entrance channel leads to the port. The port is comprised of tidal harbour and three tidal docks; there are berthing facilities on both of the Dee's banks. The overall terminal length is approximately 6 km. Depths at the berths can reach 9.3 m. (Pacific Terminal). Dockage facilities can provide complete overhaul of vessels, including destroyers.

It is clear, therefore, that the maps and descriptions are an immense repository of topographic intelligence, and provide thorough supporting information for a possible invading or occupying army, as well as for a military campaign and combat operations. However, they are broader than this in their scope, and it is possible that the maps could also have supported civil administration after a successful coup.

Where did their source information come from? Whereas the earlier city plans until the 1960s contained detailed marginal notes of sources, later maps rarely cite their sources. However, detailed research makes it clear that the Soviet cartographers used a very wide, eclectic (and sometimes inconsistent) range of sources, as well as first-hand information from spies or local informants. In the case of Scotland, official topographic maps by Ordnance Survey and commercial publishers such as Bartholomew were clearly primary sources. However, as the Soviet Union's own cartography for civilians was heavily censored and unavailable at detailed scales, the Soviet army cartographers had a natural assumption that generally available national topographic maps would be deliberately falsified or erased of significant detail too. From the launch of the Sputnik satellite in 1957 and the Zenit satellite in 1962, the Soviets were also able to use a growing quantity of satellite information, and there is good evidence for the use of satellite imagery in map compilation. Soviet military maps therefore often synthesise a number of past landscapes of different dates on the same page – for example, the Castlehill Barracks marked in green as Object 22 on the 1981 map (fig. 7.12d) were in fact demolished in the mid 1960s. Detailed reconnaissance on the ground could have spotted this detail. It is hard to find information on Russian military maps of Britain that could only have been compiled by people on the ground, but it can be found (fig. 7.12e). It would also have been hard to compile all the information in the *spravka* without at least some first-hand reconnaissance.

Mapping for nuclear war

More than any previous military technology, today's nuclear weapons make it impossible for any state or person to escape their globally catastrophic effects. Her Majesty's Naval Base Clyde is better known by its location at Faslane, the operating base for the submarines that could, if the time ever came, fire the UK's nuclear weapons. The base is the most important military site in Scotland today, employing over 3,000 service personnel and 4,000 civilians, and, at the time of writing, numbers were due to rise to over 8,000 personnel by 2020, following a decision to site the UK's entire submarine fleet at Faslane.

During the Second World War, Faslane saw active naval use, along with Rosneath to the south (fig. 7.9), and was chosen in the 1960s as the base for the Polaris missile system. Its location, on the relatively secluded, deep and easily navigable Gare Loch and Firth of Clyde allows for rapid access to the North Atlantic and beyond. During the 1960s it was also close to the American nuclear submarine base at Holy Loch, which operated from 1961 to 1992. It has been a significant part of Scotland's west-coast landscape, as well as the wider political and military landscape, for more than half a century (fig. 7.13a).

From soon after its establishment, Faslane also become the focus for demonstrations by the Campaign for Nuclear Disarmament and other anti-military groups, including Trident Ploughshares, and a permanent peace camp was established outside the base gates from 1982. During the Scottish independence referendum campaign in 2014, Faslane became a major focus for debate (fig. 7.13b). The Scottish National Party, along with the Scottish Socialist Party and the Scottish Green Party, all opposed nuclear weapons, although the Scottish National Party claimed that they would retain the base for conventional armaments. In the event, a narrow majority to the 'No' voters in 2014, followed by a House of Commons decision in 2016 to upgrade the existing Trident submarine fleet over the next twenty years, have ensured the continuation of Faslane as a nuclear base.

The recent Russian military maps of Scotland, and even these maps of Scotland's most important military base today by different British institutions, may seem a long way from John Hardyng's map of the fifteenth century, but in fact there

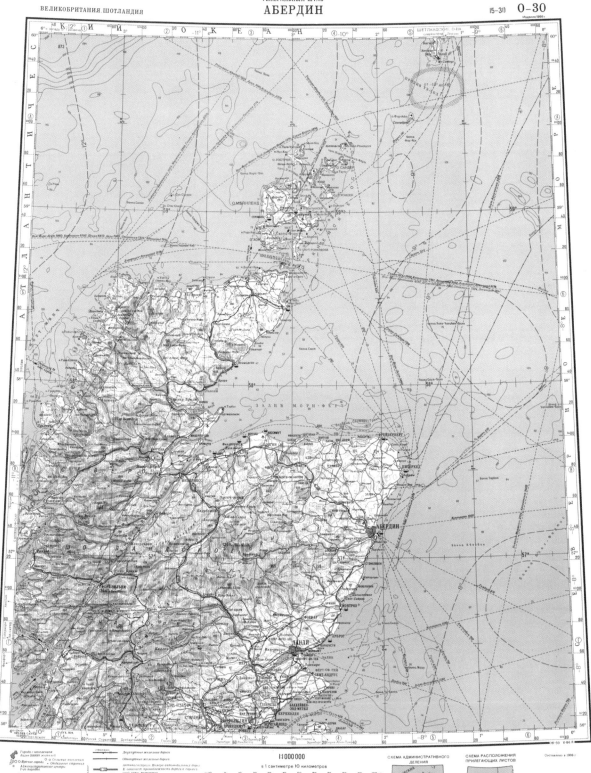

(a)

FIGURE 7.12

(a) and detail (b) (*Opposite and right*) Smaller-scale Russian military maps were for general terrain evaluation and allowing a strategic overview of wide areas. On the 1:1 million map (fig. 7.12a and b), coastal information is particularly strong, with lighthouses shown by stars, the two naval bases of Faslane and Rosyth shown with a 'molehill' symbol with two spikes on top, and distinguished from the 'boat' symbol for civilian sea ports/harbours, such as for Dundee, Arbroath and Montrose. Three different types of rocks are distinguished, according to whether they are visible at high tide, low tide or intermediate tides. Inland, apart from the significance of the terrain itself, including rivers, lochs and major areas of woodland, the maps pick out communications, particularly by rail and road. One interesting feature of the roads is the reference to a European road classification with E numbers (e.g. E120 for the A9, south of Drumochter, and E32 for the M90 south of Perth). This derived from an official European source, but was never implemented on the ground or in any British road atlases.

For the two smaller-scale sheets shown here, the Gauss–Krüger projection, as a conformal transverse cylindrical projection, allows angles to be preserved and bearings to be taken from the map (of great value for artillery and rocket fire). However, it does mean that the central meridians of each grid are not parallel to one another, explaining why the map is not rectangular, with the western and eastern sides tapering to the north, and with a curved lower margin.

(c) (*Right*) Many of the features on the 1:1 million map can be seen with greater clarity in the 1:500,000 detail of the Moray Firth. This particularly highlights the large number of military installations in this area, including Fort George (Форт Джордж) and six military airfields (civilian airfields had a grey aeroplane symbol).

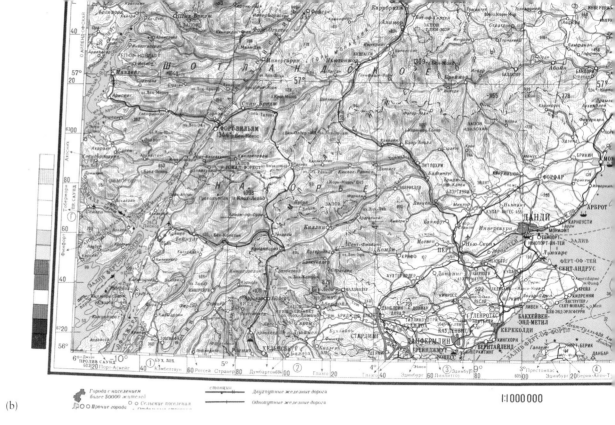

(b)

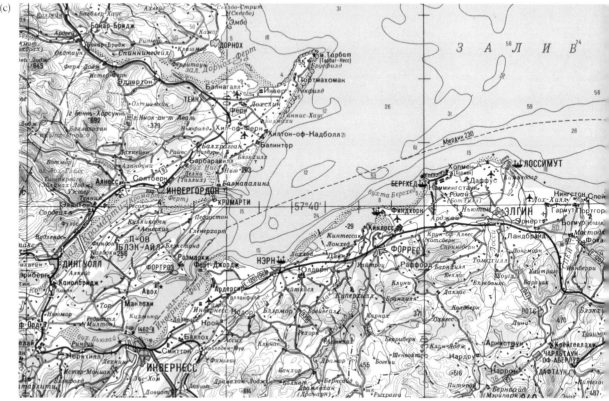

(c)

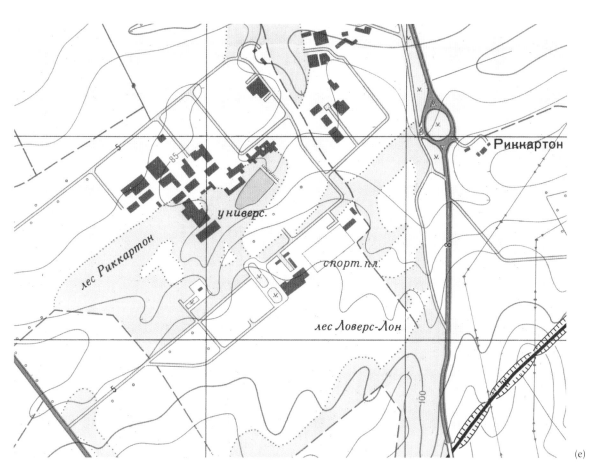

(e)

FIGURE 7.12 (continued)

(d) (*Left*) The 1:10,000 scale map of Aberdeen, from which this detail is taken, was one of ninety-three known plans of United Kingdom cities, forming a small part of around 2,000 city plans worldwide. At first glance, one of the most striking features is the colour coding of buildings, not employed by any other map-maker (brown for residential, black for industrial, green for military and purple for civil administration). The map is printed in ten colours, giving good scope for more subtle colour coding. Grey-green contour lines curve across the map at very regular 5-metre intervals, with regular spot heights on the land and depths in the harbour, as well as in all the major docks. There are also precise measurements given for the bridges over the Dee – the Victoria Bridge and Wellington Suspension Bridge have their lengths and widths accurately recorded – and between them both the width of the Dee is also measured. A set of fifty-eight 'important objects' are numbered and named in the accompanying key. Foreign names were transliterated phonetically so that the non-Russian speaker could understand them when spoken. This interesting approach, of course, theoretically enabled Russian speakers to be understood by locals, but was of no help for names on streets or road signs on the ground. The harbour is particularly packed with information. The star symbols are for lights, ГСМ for fuel and lubricants, and each of the docks is named with depths given. The *spravka* (accompanying geographical description of the city) also describes several of the named dock features: 'There are two docks (obj. 7, 8), three floating docks (cargo capacity 600, 700 and 5350 tons), two slipways (cargo capacity 600 and 900 tons), 21 shore cranes (lifting capacity of one 80, others – from 1.5 to 15 tons), coal storage facilities and trading warehouses (obj. 45), fuel and lubricant storages (obj. 43, 44).'

(e) (*Above*) It is rare to find real evidence for features that could only have been recorded by ground reconnaissance, but in this 1:10,000 detail of Riccarton Campus at Heriot-Watt University, Edinburgh, an area of recent development in the 1970s, the widths of the minor roads are recorded here and further north leading to Hermiston Gait. The first phase of the academic buildings opened in 1974, and the Russian map captures the situation a few years later.

Sources: (a) and (b) Советская Армия. Генеральный штаб, Великобритания, Шотландия, Абердин = Soviet Army, General Staff, *Great Britain, Scotland, Aberdeen (O-30)* (1986). (c) Советская Армия. Генеральный штаб, Великобритания, Шотландия, КЕРККОЛДИ = Soviet Army, General Staff, *Great Britain, Scotland, Kirkcaldy (O-30-B)* (1986). (d) Советская Армия. Генеральный штаб, Абердин = Soviet Army, General Staff, *Aberdeen (O-30-104)* (1981). (e) Советская Армия. Генеральный штаб, Эдинбург = Soviet Army, General Staff, *Edinburgh (N-30-6)* (1983).

(d)

are many close connections. The nature of warfare and military technologies may have changed beyond all recognition, but maps still provide a crucial role for those attacking, defending or simply living in Scotland. The maps contain topographic detail which may be useful, which may confer real or perceived advantages for those using them, and their creation has involved real expertise, not least in gathering first-hand topographic information and in draughtsmanship. Like John Hardyng's highly selective detail, these modern maps also exclude more than they reveal, and they show how malleable the same military landscape is when represented in map form. In some ways, the invisibility of military sites in the modern landscape, their lack of distinction from the surrounding rural landscape, as we see in OS *Open Map – Local* (fig. 7.13c), is significant, too. Globally, military technologies are more omnipotent and pervasive than ever before, and yet less and less easy to spot on the map. As we have demonstrated in this book, all the maps have a real power to promote particular causes, groups, states or nations – they encourage their users to think and act about the world presented before them in different ways. Their moral ambiguity, as true in the fifteenth century as in the twenty-first, allows them to be used to promote war, peace, attack or defence. Whatever the future holds for Scotland, and whatever military technologies prevail, we can be confident that maps will continue to play an important role.

FIGURE 7.13

(a) (*Right*) This enlarged inset at 1:6,250 scale on an Admiralty Chart from 2015 of the wider Gareloch and approaches, provides the perfect guide for navigating successfully into the Faslane nuclear submarine base. Although at first it looks deceptively easy, of course, the Restricted Area and Protected Area provide strong practical and administrative deterrents, before reaching the Floating Barrier, with a Barrier Gate, fortunately blocking easy maritime access.

(b) (*Opposite top*) The cartographic outline of Scotland as a symbol for national identity provides a convenient vehicle for alignment with many various causes. During the Scottish Independence Referendum in 2014, this poster by the Campaign for Nuclear Disarmament was used as a campaigning graphic.

(b)

(c) (*Right*) From the late twentieth century, the vast majority of maps have been produced using digital techniques. *OS Open Map – Local* is part of the freely available OS Opendata range of map datasets, and shows generalised thematic layers, including buildings, roads, sites, railways, hydrology, coastlines, woodland and names. It is updated every six months, and intended to be viewed at 1:10,000 scale.

Within the last twenty years, there has been a growing realisation, with the widespread availability of high-resolution satellite imagery and excellent, detailed mapping from many civilian and military sources beyond the control of any national government, that attempts to censor military sites from maps are both pointless and irrelevant. Faslane appears here as nothing more than an accessible small port, with scattered woodland and open space nearby, close to good leisure facilities in nearby Garelochhead. That said, although many formerly censored military sites in Scotland have made their debut in the last ten years on Ordnance Survey maps, censorship is still applied to some highly sensitive sites, which are often deliberately blurred or low resolution on Google and Bing satellite imagery.

Sources: (a) Hydrographic Office, *Gareloch, Admiralty Chart 2000* (2015). © British Crown Copyright, 2015. All rights reserved. (b) Scottish Campaign for Nuclear Disarmament, *Scotland: No Place for Nuclear Weapons* (2014). Courtesy of Scottish CND. (c) Ordnance Survey, *OS Open Map – Local, detail of Faslane Port* (2018). Contains OS data © Crown copyright and database right 2018.

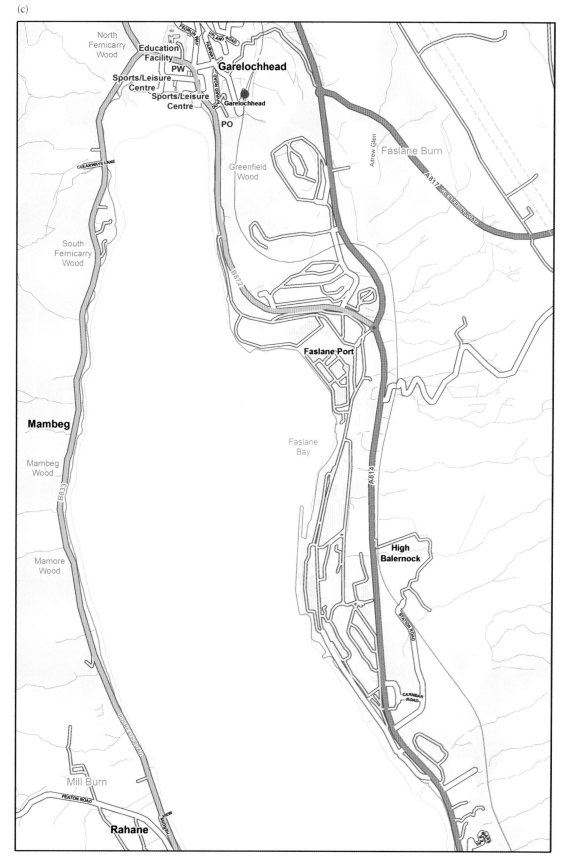

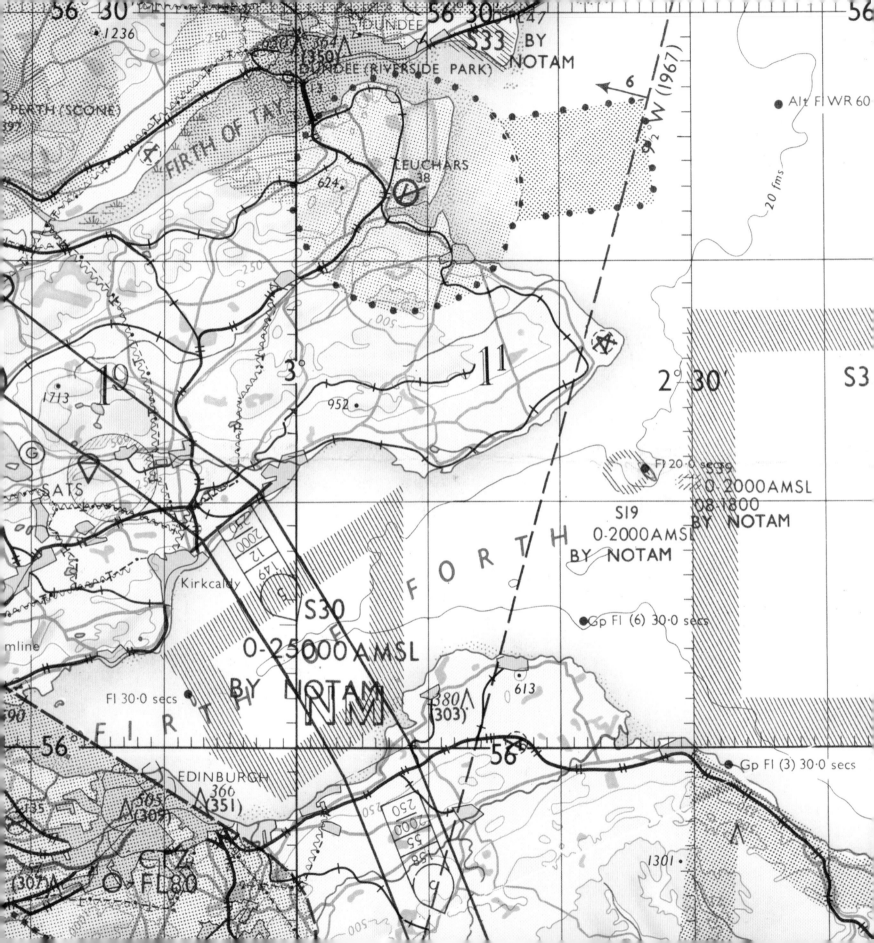

GUIDE TO SOURCES AND FURTHER READING

Chapter 1

For early maps of Scotland, the fullest general bibliographical record remains D. G. Moir (ed.), *The Early Maps of Scotland*, 2 vols (Edinburgh: Royal Scottish Geographical Society, 1973 and 1983). Also useful is J. N. Moore, *The Historical Cartography of Scotland*, O'Dell Memorial Monograph No. 24 (Aberdeen: University of Aberdeen, Department of Geography, 1991). The fullest carto-bibliography of Board of Ordnance and related military maps of Scotland can be found in C. J. Anderson, 'Constructing the military landscape: the Board of Ordnance maps and plans of Scotland, 1689–1815', unpublished University of Edinburgh doctoral thesis, 2010 (http://hdl.handle.net/1842/4598), Vol 2.

The National Library of Scotland's *Map Images* website (https://maps.nls.uk) provides access to over 200,000 high-resolution images of maps, including many which are illustrated in this book. The website includes access to all the Board of Ordnance and Wade military maps, as well as the Roy Military Survey of Scotland (1747–55) (https://maps.nls.uk/military/) along with supporting contextual and biographical information. Edinburgh University's *Charting the Nation* website (http://www.chartingthenation.lib.ed.ac.uk/) provides some overlapping coverage, as well as other maps of Scotland in the period 1550–1740. For archaeological and architectural history of specific fortifications and maritime heritage, consult the CANMORE website (https://canmore.org.uk/).

This book forms part of a series of volumes on Scottish mapping, published by Birlinn in association with the National Library of Scotland. C. Fleet, J. M. Wilkes and C. W. J. Withers' *Scotland: Mapping the Nation* (Edinburgh: Birlinn in association with the National Library of Scotland, 2011) provides a general thematic overview, with a chapter devoted to military mapping. C. Fleet, J. M. Wilkes and C. W. J. Withers' *Scotland: Mapping the Islands* (Edinburgh: Birlinn in association with the National Library of Scotland, 2016) provides more specific thematic coverage of the mapping of islands. C. Fleet and D. MacCannell's *Edinburgh: Mapping the City* (Edinburgh: Birlinn in association with the National Library of Scotland, 2014), J. Moore's *Glasgow: Mapping the City* (Edinburgh: Birlinn, 2015) and J. Moore's *The Clyde: Mapping the River* (Edinburgh: Birlinn, 2017) adopt a chronological perspective, focusing on a selection of maps of these places.

For a detailed background and commentary on John Hardyng see A. Hiatt, 'Beyond a border: The maps of Scotland in John Hardyng's *Chronicle*' in J. Stratford (ed.), *The Lancastrian Court: Proceedings of the 2001 Harlaxton Symposium* (Donnington: Shaun Tyas, 2003), 78–94. For Dacre's fortified Anglo-Scottish border map, see M. Merriman, '"The Epystle to the Queen's Majestie" and its "Platte"', *Architectural History*, 27 (1984), 25–32. B. P. Lenman (ed.), *Military Engineers and the Development of the Early-Modern European State* (Dundee: Dundee University Press, 2013) provides a detailed, recent scholarly overview of European military engineers. See especially the chapter in this volume focusing on Scotland in the eighteenth century: C. J. Anderson, 'Cartography and conflict: The Board of Ordnance and the construction of the military landscape of Scotland', 131–52. For William Roy's Military Antiquities (online at https://maps.nls.uk/roy/antiquities/), see G. Macdonald, 'General William Roy and his "Military Antiquities of the Romans in North Britain"', *Archaeologia* 68 (1917), 161–228. On the work of Lewis Petit, see C. Fleet, 'Lewis Petit and his plans of Scottish fortifications and towns, 1714–16', *Cartographic Journal* 44 (2007), 329–41. The quote about M. de Bombelles can be found in B. Faujas Sait-Fond, *Travels in England, Scotland and the Hebrides* (London: James Ridgeway, 1799), 156. Full details of the Dundas family's connections to David Watson and their patronage of the Military Survey are in R. Hewitt, 'A family affair: The Dundas family of Arniston and the Military Survey of Scotland (1747–1755)', *Imago Mundi* 64 (2012), 60–77.

Chapter 2

Alexander Lindsay's rutter and the Nicholay map of 1583 are discussed in D.G. Moir (ed.), *The Early Maps of Scotland*, Vol. 1 (Edinburgh: Royal Scottish Geographical Society, 1973), and in greater detail in I. Adams and G. Fortune (eds), *Alexander Lindsay: A Rutter of the Scottish Seas, circa 1540* (London: National Maritime Museum, 1980).

The broader context to English map-making in this period, and its impact on Scotland, is discussed in P. Barber, 'Mapmaking in England, ca. 1470–1650', in D. Woodward (ed.), *The History of Cartography, Volume 3: Cartography in the European Renaissance* (Chicago, IL: University of Chicago Press, 2007), 1589–1669. For the connections between mapping, politics and the Military Revolution in early modern Europe, see G. Parker, *The Military Revolution 1500–1800: Military Innovation and the Rise of the West*, 2nd edn (Cambridge: Cambridge University Press, 1996), D. Buisseret (ed.), *Monarchs, Ministers and Maps: The Emergence of Cartog-*

raphy as a Tool of Government in Early Modern Europe (Chicago, IL: University of Chicago Press, 1992) and D. Buisseret, *The Mapmakers' Quest: Depicting New Worlds in Renaissance Europe* (Oxford: Oxford University Press, 2003), chapter 5.

For excellent detail on the Rough Wooings, and related military engineering and maps at this time, see M. Merriman, *The Rough Wooings: Mary Queen of Scots, 1542–1551* (East Linton: Tuckwell Press, 2000). G. Phillips, *The Anglo-Scots Wars, 1513–1550* (Woodbridge: Boydell Press, 1999) is also useful on warfare between England and Scotland at this time. For further detail on particular fortifications in Scotland, see M. Merriman, 'The fortresses in Scotland, 1547–50' in H. Colvin (ed.), *The History of the King's Works, Volume 4: 1485–1660 (Part II)* (London: HMSO, 1982), 694–726. For wider British map-making and military engineering in this period, see M. Merriman, 'Italian military engineers in Britain in the 1540s', in S. Tyacke (ed.), *English Map-Making, 1550–1650* (London: British Library, 1993), 57–67 and G. Phillips, 'Scotland in the age of the Military Revolution, 1488–1560' in E. Spiers, J. A. Crang and M. J. Strickland, *A Military History of Scotland* (Edinburgh: Edinburgh University Press, 2012), 182–208. G. Macdonald Fraser, *The Steel Bonnets* (London: Collins Harvill, 1986) is a very entertaining account of Anglo–Scottish border counties in the sixteenth century, and Border reiving. Several important military maps of Scotland, including the *Platte of Milkcastle*, which were collected by William Cecil, First Baron Burghley, are described in R. A. Skelton and J. Summerson, *A Description of the Maps and Architectural Drawings in the Collection made by William Cecil, first Baron Burghley, now at Hatfield House* ([London]: [Oxford University Press for] the Roxburghe Club, 1971).

For the history and archaeology of many features in Edinburgh Castle, including the spur, see G. Ewart and D. Gallacher, *Fortress of the Kingdom: Archaeology and Research at Edinburgh Castle*, Archaeology Report no. 7 (Edinburgh: Historic Scotland, 2014). For the map of the siege of Leith see S. Harris, 'The fortifications and siege of Leith: a further study of the map of the siege in 1560', *Proceedings of the Society of Antiquaries of Scotland* 121 (1991), 359–68. M. Merriman's 'The Platte of Castlemilk, 1547', *Transactions of the Dumfriesshire and Galloway Natural History and Antiquarian Society*, 44 (1967), 175–81, gives excellent detail on this map. C. Oman's 'The battle of Pinkie, 10 September 1547', *Archaeological Journal*, 90 (1934), 1–25 offers a detailed insight into the discovery and explanation of John Ramsay's scroll map of Pinkie. D. Caldwell's 'The battle of Pinkie' in N. Macdougall, *Scotland and War AD 79–1928* (Edinburgh, 1991), 61–94 explains the battle with reference to the maps and other written evidence.

There are many detailed works on Timothy Pont's mapping. J. C. Stone's, *The Pont Manuscript Maps of Scotland: Sixteenth-Century Origins of a Blaeu Atlas* (Tring: Map Collector Publications, 1989) provides a detailed description of all the maps, while a more recent overview can be found in J. C. Stone's, 'The Kingdom of Scotland: cartography in an age of confidence', in D. Woodward (ed.), *The History of Cartography, Volume 3: Cartography in the European Renaissance* (Chicago, IL: University of Chicago Press, 2007), 1684–92. For interpretive essays on selected themes, see I. C. Cunningham (ed.), *The Nation Survey'd: Essays on Late Sixteenth Century Scotland as Depicted by Timothy Pont* (East Linton: Tuckwell Press in association with the National Library of Scotland, 2001). All the Pont maps and texts, along with contextual essays, are available on the Pont maps website (https://maps.nls.uk/pont/).

For the work of Joan Blaeu, see the *Blaeu Atlas* website (https://maps.nls.uk/atlas/blaeu/index.html), which includes a searchable facsimile of the *Atlas* together with indexes of place names, personal names and introductory and biographical essays. A printed facsimile edition of the 1654 *Atlas Novus*, with introductory essays, is also available: *The Blaeu Atlas of Scotland* (Edinburgh: Birlinn in association with the National Library of Scotland, 2006).

The Cromwellian fortifications in Scotland are described in A. A. Tait, 'The Protectorate citadels of Scotland', *Architectural History* 8 (1965), 9–24. The wider military background of this period is discussed in C. Duffy, *Fire and Stone: The Science of Fortress Warfare, 1660–1880* (London: Greenhill, 1996). For a detailed description of Dunnottar Castle, see W. D. Simpson, *Dunnottar Castle: Historical and Descriptive*, 15th ed. (Dunecht: C. A. Pearson, 1987). A good general background to John Slezer can be found in: K. Cavers, *A Vision of Scotland: The Nation Depicted by John Slezer 1671–1717* ([Edinburgh]: HMSO [in association with] National Library of Scotland, 1993) and at https://digital.nls.uk/slezer/.

Chapters 3–5

The causes of Jacobitism and the Jacobite risings are covered extensively. The role of maps and mapping is not discussed, but the historical narrative is useful for understanding the motives behind their production. See C. Kidd, *Subverting Scotland's Past: Scottish Whig Historians and the Creation of an Anglo-British Identity, 1689–c.1830* (Cambridge: Cambridge University Press, 1993); D. Szechi, *The Jacobites: Britain and Europe 1688–1788* (Manchester: Manchester University Press, 1994); B. Lenman, *The Jacobite Risings in Britain, 1689–1746* (Dalkeith: Scottish Cultural Press, 1995); and A. I. Macinnes, 'Scottish Jacobitism: in search of a movement', in T. M. Devine and J. R. Young (eds), *Eighteenth Century Scotland: New Perspectives* (East Linton: Tuckwell Press, 1999), 70–89.

On the Board of Ordnance as a whole in Scotland, see C. J. Anderson, 'State imperatives: military mapping in Scotland, 1689–1770', *Scottish Geographical Journal* 125 (2009), 4–24. The general connections between mapping becoming militarised and the military becoming map-minded in the eighteenth century are explored in M. Edney, 'British military education, mapmaking, and military "map mindedness" in the later Enlightenment', *Cartographic Journal* 31 (1994), 14–20. For more on the Board of Ordnance as a British institution for military mapping, see D. W. Marshall, 'Military maps of the eighteenth century and the Tower of London Drawing Room', *Imago Mundi* 32 (1980), 21–44; R. W. Stewart, *The English Ordnance Office, 1585–1625: A Case Study in Bureaucracy* (London: The Royal Historical Society, 1996); and H. C. Tomlinson, *Guns and Government: The Ordnance Office under the later Stuarts* (London: Royal Historical Studies in History, 1979). The establishment of the engineers is the focus of a study by W. Porter, *History of the Corps of Royal Engineers*, Vol. 1 (London: Longmans, Green and Co., 1889).

Extensive use has been made of map evidence by historians, geographers, architects and archaeologists studying Scotland's eighteenth-century fortifications. See, for example, I. MacIvor, *Edinburgh Castle* (London: Batsford, 1993); C. Tabraham and D. Grove, *Fortress Scotland and the Jacobites* (London: B. T. Batsford / Historic Scotland, 1995); D. Gallagher and G. Ewart, *Stirling Castle Palace: The History and Archaeology of Stirling Castle Palace* (Edinburgh: Historic Scotland, Archaeological and Historical Research 2004–2008); and G. Ewart and D. Gallagher, 'The fortifications of Fort George, Ardersier, near Inverness: archaeological investigations 1990–2005', *Post Medieval Archaeology* 44:1 (2010), 105–34. Barracks are the focus in J. Douet, *British Barracks, 1600–1914: Their Architecture and Role in Society* (London: The Stationery Office for English Heritage, 1998), 1–28. A useful reference for understanding some of the features of fortification architecture is J-D. Lepage, *French Military under Louis XIV: An Illustrated History of Fortifications and Strategies* (Jefferson: McFarland & Company, 2010). For a detailed study of the assimilation of the Scots into the British Army, see V. Henshaw, *Scotland and the British Army, 1700–1750: Defending the Union* (London: Bloomsbury Academic, 2014).

For more about the military road system in eighteenth-century Scotland, see J. Mathieson, 'General Wade and his military roads in the Highlands of Scotland', *Scottish Geographical Journal* 40:4 (1924), 193–213; T. Ruddock, 'Bridges and roads in Scotland: 1400–1750' in A. Fenton and G. Stell (eds), *Loads and Roads in Scotland and Beyond: Land Transport over 6000 Years* (Edinburgh: John Donald, 1984), 67–91; W. Taylor, *The Military Roads in Scotland* (Argyll: House of Lochar,

1996); and L. Farquharson, *General Wade's Legacy: The 18th-Century Military Road System in Perthshire* (Perth: Perth and Kinross Heritage Trust, 2011).

Maps bring battle narratives to life. For studies of Scotland's eighteenth-century battlefields, see A. H. Millar, 'The battle of Glenshiel, 10th June 1719: Note upon an unpublished document in the possession of His Grace the Duke of Marlborough', *Proceedings of the Society of Antiquaries of Scotland* 17 (1882), 57–69; G. Foard, *Draft Report on Prestonpans* (prepared for Historic Scotland by the Battlefields Trust, 2005); G. Foard, *Draft Report on Sheriffmuir* (prepared for Historic Scotland by the Battlefields Trust, 2005); T. Pollard, 'Mapping mayhem: Scottish battle maps and their role in archaeological research', *Scottish Geographical Journal* 125 (2009), 25–42; and R. Woosnam-Savage, '"To gather an image whole": Some early maps and plans of Culloden' in T. Pollard (ed.), *Culloden: The History and Archaeology of the Last Clan Battle* (Barnsley: Pen and Sword, 2009). The Jacobite Rising of 1745 has excited much attention as an episode in Scotland's turbulent history. For more about the reasons behind the rising, and its successes and losses, see M. Hook and W. Ross, *The 'Forty-Five: The Last Jacobite Rebellion* (Edinburgh: HMSO, 1995); and C. Duffy, *The '45* (London: Cassell, 2003).

The Military Survey of Scotland has been the focus of much study. See, for example, D. G. Moir (ed.), *The Early Maps of Scotland*, Vol. 1 (Edinburgh: Royal Scottish Geographical Society, 1973), 103–13; and R. A. Skelton, 'The Military Survey of Scotland 1747–1755', *Scottish Geographical Magazine* 83 (1967), 5–16. This was reprinted as R. A. Skelton, *The Military Survey of Scotland, 1747–1755* (Edinburgh: Royal Scottish Geographical Society, Special Publication Number 1, 1967). See also G. Whittington and A. J. S. Gibson, *The Military Survey of Scotland 1747–1755: A Critique*, Historical Geography Research Series, No. 18 (Norwich: Geo Books, 1986). The best study of William Roy and the Military Survey is provided by Y. O'Donoghue, *William Roy 1726–1790: Pioneer of the Ordnance Survey* (London: British Library, 1977). Roy's 'Great Map' is available as a facsimile edition: [W. Roy], *The Great Map: The Military Survey of Scotland, 1747–1755*; with introductory essays by Y. Hodson, C. Tabraham and C. W. J. Withers (Edinburgh: Birlinn in association with the National Library of Scotland, 2007). The map is available for consultation electronically via the Map Division of the National Library of Scotland: https://www.nls.uk/maps/roy/index.html. This link has accompanying essays on the maps, their content, and on William Roy.

A good general background to coastal defences, martellos and other artillery fortifications can be found in R. Morris and G. J. Barclay, 'The fixed defences of the Forth in the Revolutionary and Napoleonic Wars, 1779–1815', *Tayside and Fife Archaeological Journal* 23 (2017), 109–33.

Chapter 6

The most detailed and authoritative account of prisoners of war in Scotland can be read in I. MacDougall, *All Men are Brethren: Prisoners of War in Scotland, 1803–1814* (Edinburgh: John Donald, 2008). For the specific role of Edinburgh Castle, and the vaults beneath the Great Hall and Crown Square, see C. Tabraham, 'The prisons of war in Edinburgh Castle', *Book of the Old Edinburgh Club*, New series, Vol. 7, (2008) 53–69. A. Saunders, *Fortress Britain: Artillery Fortification in the British Isles and Ireland* (Liphook, Hants: Beaufort, 1989) provides good detail on the changing engineering principles behind British military defences over time.

There are several good books on different aspects of Ordnance Survey's history. R. Hewitt's *Map of a Nation: A Biography of the Ordnance Survey* (London: Granta, 2010) provides an accessible and well-researched narrative from the work of William Roy through to the mid nineteenth century. Ordnance Survey's connections to military mapping and the War Office are well covered in W. A. Seymour's *A History of the Ordnance Survey* (Folkestone: Dawson, 1980), and in passing in R. Oliver, *The Ordnance Survey in the Nineteenth Century: Maps, Money and the Growth of Government* (London: Charles Close Society, 2014).

For the broader cartographic context to mapping during the First World War, see P. Chasseaud, *Mapping the First World War* (London: Imperial War Museum, 2013). C. M. M. Macdonald and E. W. McFarland, *Scotland and the Great War* (East Linton: Tuckwell Press, 1999) provides useful detailed essays on several aspects of the impact of the First World War on Scotland. For more specific detail on the development of the Rosyth naval dockyard, see W. Burt, *Rosyth Dockyard and Naval Base through Time* (Stroud: Amberley Publishing, 2016), B. Lenman, *From Esk to Tweed: Harbours, Ships and Men of the East Coast of Scotland* (Glasgow: Blackie, 1975) and R. Paxton and J. Shipway, *Civil Engineering Heritage: Scotland – Lowlands and Borders* (London: Thomas Telford, 2007).

Chapter 7

For a brief general overview of twentieth-century military mapping, consult J. Black, 'War and cartography' in T. Harper (ed.), *Maps and the 20th Century: Drawing the Line* (London: British Library, 2016), 31–65 and P. Collier and M. Nolan, 'Military mapping by Great Britain' in M. Monmonier (ed.), *The History of Cartography, Volume 6: Cartography in the Twentieth Century* (Chicago, IL: University of Chicago Press, 2015), 894–904. A detailed background to British military mapping in the interwar period can be found in Y. Hodson, 'MacLeod, MI4, and the Directorate of Military Survey 1919–1943', *Cartographic Journal* 38 (2001), 155–75.

L. Taylor's *Luftwaffe over Scotland* (Dunbeath: Whittles Publishing, 2010), is a thorough guide to German air attacks on Scotland, while S. Allen's *Commando Country* (Edinburgh: NMS Enterprises, 2007) provides well-researched descriptions of special service military training grounds in Scotland. G. Barclay's *If Hitler Comes: Preparing for Invasion: Scotland 1940* (Edinburgh: Birlinn, 2013) provides a detailed account of the anti-invasion defences constructed in Scotland in 1940–41 and what survives today. M. M. Evans' *Invasion! Operation Sealion, 1940* (Harlow: Longman, 2004) provides detailed coverage of the German preparations for invading Britain in 1940. For German military mapping see J. Neumann, 'Military mapping by Germany' in M. Monmonier (ed.), *The History of Cartography, Volume 6: Cartography in the Twentieth Century* (Chicago, IL: University of Chicago Press, 2015), 909–21. For naval bases, harbours and the navy in Scotland see B. Lavery, *Shield of Empire: The Royal Navy and Scotland* (Edinburgh: Birlinn, 2007).

For the early development of aeronautical charts in the United Kingdom, see T. R. Nicholson, 'An introduction to the Ordnance Survey aviation maps of Great Britain, 1925–39', *Sheetlines* 23 (1988), 5–18. 'Aeronautical charts', chapter 14 in C. R. Perkins and R. B. Parry, *Mapping the United Kingdom* (London: Bowker-Saur, c.1996) provides a good overview of British air mapping up to the 1990s. For a more recent general history, see R. E. Ehrenberg, 'Aeronautical chart' in M. Monmonier (ed.), *The History of Cartography, Volume 6: Cartography in the Twentieth Century* (Chicago, IL: University of Chicago Press, 2015), 22–9. R. Blake, 'Charting the aeronautical landscape', *Sheetlines* 99 (April 2014), 19–39 (https://www.charlesclosesociety.org/files/Issue99page19.pdf) reports on a detailed survey of Ordnance Survey maps and their limited depiction of airfields for security reasons.

Russian military mapping of the world is examined in detail by J. Davies and A. J. Kent in *The Red Atlas: How the Soviet Union Secretly Mapped the World* (Chicago, IL: University of Chicago Press, 2017). J. Davies' website https://sovietmaps.com gathers many other resources together, including various earlier papers: 'Uncle Joe knew where you lived – Part 1', *Sheetlines* 72 (April 2005); Part 2 in *Sheetlines* 73 (August 2005). D. Watt, 'Soviet military mapping', *Sheetlines* 74 (December 2005), describes the history of the Russian Military Topographic Directorate (VTU) from 1812 to the present day. J. Cruickshank, 'Military mapping by Russia and the Soviet Union' in M. Monmonier (ed.), *The History of Cartography, Volume 6: Cartography in the Twentieth Century* (Chicago, IL: University of Chicago Press, 2015), 932–42 provides a good recent summary.

INDEX

Items in **bold** indicate figures.

Aberchalder 98
Aberdeen vii, 3, 13, 99, 100, 107, 109, 131, 155, 210, 215
    Plan of Cumberland Fort at Aberdeen (1746) **13**
    Soviet Army, General Staff, *Aberdin (O-30-104)* (1981) **214**
Adair, John 41, 109
    The Mapp of Straithern, Stormount, and Carsof Gourie, with the Rivers Tay and Iern (c.1720) **41**
Adams, John
    Drawing of a circumferentor (1799) **123**
Admiralty *see* Hydrographic Office
aerial reconnaissance *see* reconnaissance
aerodromes *see* airfields/aerodromes
aeronautical charts 192, 203, 209
    British Army of the Rhine, *Ascent Topographic Low Flying Chart: Europe 1:500.000. Sheet 7. United Kingdom, North* (1967) **208–9**, 218
Ainslie, John
    Old and New Town of Edinburgh and Leith with the proposed docks (1804) **147**
air photos 197, 199, 203, 206–7
    German Army, oblique air photo, Blick von Queensferry über die Forthbrücke... (1940) **197**
    German Army, oblique air photo, Borrowstouness (Bo'ness) (1940) **197**
    Ordnance Survey, *Air Photo Mosaic 26/75 S.W.* (1946) **206**
    Ordnance Survey, *Air Photo Mosaic 26/75 S.W.* (1950) **207**
airfields/aerodromes 14, 185–86, 190, 193, 195, 209, 213
    Ordnance Survey, Scotland, One-Inch to the mile, Popular edition, *Sheet 68, Firth of Forth* (revised 1923–6, published 1928) **190**
    John Bartholomew & Son, Half-Inch to the mile, Scotland, *Sheet 8 Forth* (1934) **191**
    Quarter-Inch to the mile. *Sheet 3 – The Forth & Tay* (1939) 21, **194–195**
    War Office / Geographical Section, General Staff, Great Britain, 1:25,000 GSGS.3906, *Sheet 29/68 S.W.* (1941) **186**
    Ordnance Survey, Scotland, One-Inch to the mile, Popular edition with National Grid, *Sheet 68, Firth of Forth* (1924 with later revisions 1945) **191**
    Ordnance Survey, Great Britain, 1:25,000 Provisional edition, *Sheet NS56* (1958) **187**
    British Army of the Rhine, *Ascent Topographic Low Flying Chart: Europe 1:500.000. Sheet 7, United Kingdom, North* (1967) **208–9**, 218
    Soviet Army, General Staff, *Great Britain, Scotland, Kirkcaldy (O-30-B)* (1986) **213**
American Revolutionary War (1775–83) 56, 131, 145, 156, 157, 158, 163
Anderson, Sir John 196
Anglo–Dutch Wars (1650s–1670s) 38, 151
Anglo–Scottish border 4, 35, 36
Anne, Queen 56, 57
Arbroath 171
    William Gravatt?, *Sketch of the Harbour, Town & Battery of Arbroath* (c.1795) **155**
Arcano, Archangelo 30
archaeology 11, 29
Archer, John
    A Survey of the Road made by Coll Rich's & Genl Guise's regts between Fort William & the head of King-loch Leven (1750) **136–37**
Ardersier *see* Fort George (Ardersier)
Ardhallow 173

Opposite. John Hardyng, 'Map of southern Scotland', from the *Chronicle* of John Hardyng (c.1470s). Courtesy of the Bodleian Library, University of Oxford.

Arrowsmith, Aaron 123, 145, 156
Avery, Joseph
    This Plan Containing Lochness, Lochoyoch, Lochlochey, & all the Rivers and Strips of Water... (1727) **90–91**
    A Plan of the Country where the New Intended Road is to be made from the Barack at Ruthven in Badenoth to Invercall in Brae Marr... (1735) **81**
aviation *see* airfields/aerodromes
Augustus, William (Duke of Cumberland) 13, 84, 99, 102, 104, 107, 109, 111, 113, 116, 118, 124, 126, 141

Balgillo Fort 30
Balquhidder 68, 135
barrack forts *see* barracks
barracks 7, 12, 18–19, 38, 42–43, 53, 57, 63–64, 67–68, 70, 76, 84, 86, 88, 92, 100, 107, 116, 118, 124, 126, 129, 131, 142, 146–47, 151, 157, 159, 161, 164–67, 211
    John Henri Bastide, *A Prospect of that Part of the Land and Sea adjacent to ye Barrack to be Built in Glen Elg* (1720) **53**
    Dugal Campbell, *Plan, Elevation and Sections of additional Barracks of Timber ... to be built at Fortwilliam ...* (1746) **117**
    William Skinner and Charles Tarrant, *Plans Section & Elevation of the Barrack at Cargarff* (c.1750) **128–129**
    Ordnance Survey, 25-inch to the mile, *Renfrew, Sheet XII.3* (surveyed 1858, published 1864) **166**
    Ordnance Survey, Large-Scale Town Plan, *Paisley, Sheet XII.3.11* (surveyed 1858, published 1864) **167**
Barry Buddon Ness 185
    War Office / Geographical Section General Staff, *General plan of Barry Links* (1939) **188**
    Hydrographic Office, *River Tay. Admiralty Chart 1481* (revised 1941) **189**
Bartholomew, John & Son 211
    Half–Inch to the mile, Scotland, *Sheet 8 Forth* (1934) **191**
Bastide, John Henri 70
    The Roads between Innersnait Ruthvan of Badenock Kiliwhiman and Fort William... (1718) **68–69, 78–79**
    A Draught of Innersnait, in the Highlands of North Brittain... (1719) **68**
    A Plan of the Field of Battle that was fought on ye 10th of Iune 1719, at the Pass of Glenshiels in Kintail... (1719) **72–73**
    A Prospect of that Part of the Land and Sea adjacent to ye Barrack to be Built in Glen Elg (1720) **53**
    A General Survey of Inverness, & the Country adjacent to the Foot of Loch-Ness; West Prospect of Inverness (c.1725) **85**
Beaton, Cardinal David 25
Beaton, Cardinal James 25
Beckmann, Martin (First Engineer of Great Britain) 15
Bernera 53, 57, 67, 70, 126, 130
    Andrews Jelfe, *Killewhiman, Inversnait, Ruthven of Badenoch, Bernera* (1719) **66**
    John Henri Bastide, *A Prospect of that Part of the Land and Sea adjacent to ye Barrack to be Built in Glen Elg* (1720) **53**
Blackness Castle 42, 56, 147, 154
    Theodore Dury, *Plan of the Castle of Blackness* (c.1690) **42**
Blaeu, Joan 38, 97, 109, 118
    Timothy Pont / Joan Blaeu, *Lidalia vel Lidisdalia regio, Lidisdail* (1654) **36**
Blair Castle 68, 100
Blakeney, William 99, 104
blockades 164, 173

Bo'ness
    Oblique air photo – Borrowstouness (Bo'ness) (1940) **197**
Board of Ordnance
    cartographic styles/conventions 7, 54, 73, 85, 119, 123, 141, 168
        see also map colour/colouring
    Drawing Room, Edinburgh Castle 118
    Drawing Room, Tower of London 54, 56, 118, 164
    history and functions 38, 52–54, 56
    transfer of responsibilities to Ordnance Survey and GSGS 168
Bonnie Prince Charlie *see* Stuart, Prince Charles Edward (Bonnie Prince Charlie / The Young Pretender)
Braemar Castle 126
    Repairs of Braemarr Castle (1749) **128**
Bressay Sound 145
    Andrew Frazer, *Plan of the Bay called Brassa Sound* (1786) **151**
    Dépôt Général de la Marine, *Carte des Îles Shetland* (1806) **156**
British Army of the Rhine 203
    Ascent Topographic Low Flying Chart: Europe 1:500.000. Sheet 7, United Kingdom, North (1967) **208–9, 218**
Broughty Castle 30
Bryce, Alexander 109
Bullock, Henry 36
    West Border Land (1552) **37**

Callander 95, 130
Campaign for Nuclear Disarmament *see* Scottish Campaign for Nuclear Disarmament
Campbell, Dugal 116
    Plan, Elevation and Sections of additional Barracks of Timber ... to be built at Fortwilliam ... (1746) **117**
Campbell, John (fourth Earl of Loudoun) 118
Cape Wrath 209
Cardwell, Edward (Secretary for War) 168
Castle Duart 19, 38, 44–45, 56
    The ground platt [plan] of Dowart [Duart] Castle in mull Island (1653) **44–45**
    Lewis Petit, *Plan of Castle Dwart in the Island of Moll...* (1714) **19**
Castle Tioram 56, 57, 126
Castlemilk 31
    Thomas Petit?, *The platte of Milkcastle* (1547) **32**
Caulfeild, William 81, 92, 96, 130–31, 135
Cecil, William (Lord Burghley) 31, 36
censorship 14, 168–69, 175, 185–86, 190, 193, 202, 206–7, 211, 217
Chanonry Point
    Andrew Fraser, Sketch of the Ground around Fort George (1785) **169**
    Ordnance Survey, Six-Inch to the mile, *Ross and Cromarty, Sheet XC* (revised 1904, published 1907) **169**
Charles I, King 41
Charles II, King 42, 45, 47
Churchill, Winston 163
censorship 14, 168–69, 175, 185, 186, 190, 193, 202, 206–7, 216–17
circumferentor 119
    John Adams, drawing of a circumferentor (1799) **123**
citadels 38, 43, 51, 64, 86, 124
clans 19, 36, 44, 50, 56–57, 70, 77, 80, 83, 84, 92, 113
    John Manson, Map of part of Scotland, showing clans that rebelled in 1715 (1731) **82–83**

INDEX

Cloch Point 173, 179
    War Office / Geographical Section General Staff, *Clyde, Cloch Point* (1918) 178–79
Close, Albert
    The naval war in the North Sea... (1922) 16–17
Close, Charles 168
Clyde, Firth of 3, 14, 77, 154, 172–73 *see also* Cloch Point / Faslane
Clydebank 193, 196, 202
    Department of Health, Burgh of Clydebank, Map of bomb hits on Parkhall (1941) 202
Cobbett, Ralph 44
Coehorn, Menno van 111
Colebrooke, Pawlett William
    Plan of the River or Firth of Forth (1782) 148–49
Collins, Greenvile 27
    Map of Leith from the North (1693) 43
colour/colouring *see* map colour/colouring
Columbine, Edward Henry 156
Cooper, Richard
    A map of His Majesty's Roads from Edinburgh to Inverness, Fort Augustus & Fort William (c.1742) 97–98
Cope, Sir John 98–100, 103
Corgarff Castle 126
    Old Plan of Corgarff Castle; Repairs; Repairs of Braemarr Castle (1749) 128
    William Skinner and Charles Tarrant, *Plans Section & Elevation of the Barrack at Corgarff* (c.1750) 129
Corrieyairack Pass 68, 92, 98, 100
Covenanters 29, 39, 41, 47
Crieff 92, 96. 98, 100, 107, 109
Cromwell, Oliver vi, 14, 20, 24, 38, 42–45, 47, 51, 57, 64, 85, 124, 151
Culloden, battle of (1746) 13, 80–81, 107, 109–113, 116, 118, 119, 141
    Henry Schultz, *Plan of the Battle of Collodden* (1746) 109
    Jasper Leigh Jones, *A Plan of ye Battle of Colloden...* (1746) 110–11
    Plan exact de la disposition des Troupes Ecossoises... et de Celle des Troupes Angloises a la Bataille de Culloden... 1746 (1748) ii, 112–13
Cumberland, Duke of *see* Augustus, William (Duke of Cumberland)
Cunningham, William
    Map of the battle of Falkirk, 17 January 1746 (1746) 104–5

Dacre, Christopher
    Plan and bird's-eye view of an 'Inskonce' (sconce or small fort) for the defence of the English border with Scotland (c.1583-4) 4–5
Dalwhinnie 92, 100
Debatable Land vi, 24, 35–36
    Henry Bullock, *West Border Land* (1552) 37
    Timothy Pont / Joan Blaeu, *Lidalia vel Lidisdalia regio, Lidisdail* (1654) 36
Debbeig, Hugh 118, 123
Dépôt Général de la Marine
    Carte des Îles Shetland (1806) 156
Dere Street 11
Disarming Act (1716) 80, 84, 96
Disarming Act (1746) 6, 84, 116
Donibristle 190, 195
    John Bartholomew & Son, Half-Inch to the mile, Scotland, Sheet 8 Forth (1934) 191
    Ordnance Survey, Scotland, One-Inch to the mile, Popular edition, Sheet 68, *Firth of Forth* (revised 1923–6, published 1928) 190
    Ordnance Survey, Scotland, One-Inch to the mile, Popular edition with National Grid, Sheet 68, *Firth of Forth* (1924 with later revisions 1945) 191
Douglas, Archibald (Earl of Moray) 39
Doune Castle 102
Drawing Rooms *see* Board of Ordnance
Drem 195
Drummond Castle
    John Adair, *The Mapp of Straithern, Stormount, and Cars of Gourie, with the Rivers Tay and Iern* (c.1720) 41
Duart Castle *see* Castle Duart
Dudley, John (Duke of Northumberland) 23
Duffus Castle 39
    Timothy Pont, Detail of Duffus Castle and Spynie Palace, [Gordon] 23 (c.1583–1614) 39
Dumaresq, John
    The Roads between Innersnait Ruthvan of Badenock Kiliwhiman and Fort William... (1718) 68–69, 78–79
    A Draught of Innersnait, in the Highlands of North Brittain, nere the Head of Loch Lomend... (1719) 68
Dumbarton Castle 3, 23, 50, 92, 126, 154
    Paul Sandby, *Plan of the Castle of Dunbarton* (c.1747) 127
Dunbar 2, 30, 38, 99, 100, 155
Dundas (family) 20
Dundas, David 20, 118, 157
Dundas, Henry (1st Viscount Melville) 20, 149
Dundas, William 20, 118
Dundee 3, 38, 192, 196, 213
    German Army. Schottland 1:50,000, *Blatt Nr 64 – Dundee und St Andrews* (1941) 201
Dundrennan 209
Dunkeld 50, 92, 96, 100, 119
    William Roy, Military Survey of Scotland (1747–1752) 120–21
Dunvegan Castle 45
Dunnottar 38
    John Slezer, *Plan of Dunotter* (1675) 46
Dury, Theodore 7, 42, 56, 60, 61, 76
    Plan of the Castle of Blackness (c.1690) 42
    A Plan of Sterling Castle...; A profile of Elphinstons Tower and French Spur at Sterling Castle (c.1708) 60

Edgar, William 109
    Description of the River Forth above Stirling (1746) 102
Edinburgh 3, 6, 24, 27, 28–29, 30, 31, 34, 35, 38, 41, 50, 54, 56, 58, 61, 67, 70, 75, 92, 95, 99, 100, 104, 118, 126, 130, 131, 132, 145, 146–47, 157, 163, 164, 165, 192, 196, 215
    Rowland Johnson, Siege of Edinburgh Castle (1577) 28
    James Gordon, *Edinodunensis Tabulam...* (1646) 29
    John Slezer, *A coloured bird's-eye view of Edinburgh Castle, showing the projected outworks* (c.1690) 54–55
    Talbot Edwards, *A Plan of Edinburgh Castle* (1710) 61
    Thomas Moore, Andrews Jelfe, No. 4, Edinburgh Castle (1719) 75
    Thomas Moore, Andrews Jelfe, No. 1, Edinburgh Castle (1719) 76
    A Plan of part of Edinburgh Castle (1735) 93
    John Lambertus Romer, *A Plan of Edinburgh Castle* (1737) 61
    David Watson, *Plan & Section of the Powder Magazine as it is at present, containing 684 Barrills of Powder* (1747) 132

David Watson, *Plan & Section of the Powder Magazine with the alterations propos'd* (1747) **132**
Charles Tarrant, *Plans and Sections of the Several Vaults and Floors... in the Edinburgh Castle 1754* (1754) **162–63**
William Skinner, *General Plan, Sections and Elevations of the Powder Magazine and Storehouses, Built in Edinburgh Castle Ano: 1753 & 1754* (1754) **133**
Henry Rudyard, *Plan of Part of Edinburgh Castle Showing the Proposed Situation for a Pipe & Cistern* (1794) **160–61**
Edward I, King 3
Edward VI, King 24
Edwards, Talbot 15, 61, 63
A Plan of Edinburgh Castle (1710) **61**
Eildon Hills 11
William Roy, *Plan of the Environs of the Eildon Hills on the South Bank of the Tweed...* (1793) **10**
Eilean Donan 56, 57, 70, 71
Lewis Petit, *Plann of the Castle of Island Dounan; Profile of the Front of the Castle of Island Dounan* (1714) **71**
Elder, John 15, 30
Elphinstone, John 109
A Plan of the Grounds Adjacent to Fort William (1748) **63**
A New Map of North Britain... (1746) 80, **104**
A New & Correct Mercator's Map of North Britain (1746) **108**
Erskine, John (Earl of Mar) 57, 65
Esk, River 31, 35, 36
evacuation 7, 192, 196
Evacuation Map of Scotland, *The Scotsman* (1939) **196**
Falkirk, battle of (1746) 81, 99, 104, 107, 116
William Cunningham, *Map of the battle of Falkirk, 17 January 1746* (1746) **104–5**
J. Millan, *A Plan of the Victory of Falkirk Muir Fought the Afternoon of January 16 1746* (1746) **106**
Fall, Captain William 155
Faslane 211, 213, 216
Hydrographic Office, *Gareloch. Admiralty Chart 2000* (2015) **216**
Scottish Campaign for Nuclear Disarmament, *Scotland: no place for nuclear weapons* (2014) **217**
Ordnance Survey, *OS Open Map – Local*, detail of Faslane Port (2018) **217**
Faujas de Saint-Fond, Barthélemy 15
'Fifteen (1715 Jacobite Rising) *see* Jacobite risings
First World War 16, 154, 163, 167, 172–73
War Office / Geographical Section General Staff, *Clyde, Cloch Point* (1918) **178–79**
War Office / Geographical Section General Staff, *Plan of Special Survey Inchmickery, Firth of Forth* (1918) **176–77**
Hydrographic Office, *British Isles, English Channel and North Sea. Wreck Chart in 7 sheets. Sheet II. D20* (1919) **180**
Hydrographic Office, *Waters surrounding the British Islands. Mined areas and safe channels* (1919) **181**
Albert Close, *The naval war in the North Sea...* (1922) **16–17**
Hydrographic Office, *Admiralty Chart F.86: Approaches to Rosyth Dockyard* (1944) **174–75**
Fletcher, Andrew (Lord Milton) 102
Forbes, Duncan, of Culloden 96
Fort Augustus 67, 68, 84, 88, 89, 91, 92, 126

John Dumaresq, John Henri Bastide, *The Roads between Innersnait Ruthvan of Badenock Kiliwhiman and Fort William...* (1718) **68–69, 78–79**
Andrews Jelfe, *Killewhiman, Inversnait, Ruthven of Badenoch, Bernera* (1719) **66**
John Lambertus Romer, George Wade, *A plan of the intended Fortress with the Situation of Killiwhymen* (1729) **88**
John Lambertus Romer, *Fort William; Fort Augustus; Fort George, Inverness* (c.1729–46) **114–15**
John Lambertus Romer, *The Plan of Fort Augustus, in the Highlands of Scotland...* (c.1742) **89**
Fort Charlotte 145, 146, 151, 156
Andrew Frazer, *Plan of Fort Charlotte in March 1781* (1781) **150**
Andrew Frazer, *Plan of the Bay called Brassa Sound* (1786) **151**
Dépôt Général de la Marine, *Carte des Îles Shetland* (1806) **156**
Fort George (Ardersier) 118, 119, 129, 130, 131, 141, 142, 144, 145, 168, 169, 213
William Skinner, Charles Tarrant, *Plan of the Point of Land at Arderseer with the Design'd Fort as Trac'd Thereon* (1748) **140–41**
William Skinner, Charles Tarrant, *A Plan of Fort George, North Britain, Shewing how Far Executed* (1753) **142**
William Skinner, Charles Tarrant, Fort George, Ardersier – sections (1753) **143**
Andrew Frazer, *Present Road leading Through the Garrison [and] Proposed Road to the Ferry* (1785) **144**
Andrew Fraser, *Sketch of the Ground around Fort George* (1785) **169**
Ordnance Survey, Six-Inch to the mile, *Ross and Cromarty*, Sheet XC (revised 1904, published 1907) **169**
Fort George (Inverness) 18, 84, 86, 107, 116, 124
John Lambertus Romer, *Fort William; Fort Augustus; Fort George, Inverness* (c.1729–46) **114–15**
John Lambertus Romer, *A Plan of Fort George, & part of the Town of Inverness, with proper Sections relating to the Fort* (1732) **86–87**
Charles Tarrant, *Plan of Fort George at Inverness, Shewing it's present Condition* (1750) **18**
Fort Matilda 173
Fort William 7, 15, 19, 50, 51, 56, 63, 68, 84, 85, 91, 92, 107, 95, 98, 116, 119, 124, 126, 130, 137
John Slezer, *Innerlochie or Obrian Fort, in Lochabor; Desine for Innerlochie* (1689) **51**
John Dumaresq, John Henri Bastide, *The Roads between Innersnait Ruthvan of Badenock Kiliwhiman and Fort William...* (1718) **68–69, 78–79**
John Lambertus Romer, *Fort William; Fort Augustus; Fort George, Inverness* (c.1729–46) **114–15**
George Wade, *A Plan of Fort William in the Shire of Inverness* (1736) **62**
Dugal Campbell, *Plan, Elevation and Sections of additional Barracks of Timber ... to be built at Fortwilliam ...* (1746) **117**
John Elphinstone, *A Plan of the Grounds Adjacent to Fort William* (1748) **63**
Forth Bridge
Oblique air photo – Blick von Queensferry über die Forthbrücke... (1940) **197**
Forth, Firth of 3, 31, 34, 56, 77, 131, 149, 154, 172, 177, 190–91, 195, 197, 199, 209
Pawlett William Colebrooke, *Plan of the River or Firth of Forth* (1782) **148–49**
Ordnance Survey, Scotland, 1:253,440. Sheet 3 – *The Forth & Tay* (1939) **21, 194**

INDEX

[Landing Beaches], Quarter-Inch to the mile with overprint (1940) **198–99**
British Army of the Rhine, *Ascent Topographic Low Flying Chart: Europe 1:500.000. Sheet 7, United Kingdom, North* (1967) **218**
Forth, River 58, 102
'Forty-five (1745 Jacobite Rising) *see* Jacobite risings
'Foul Burn' (Edinburgh) 165
Fraser, Simon (Lord Lovat) 80
Frazer, Andrew 144, 146, 151
    Plan of Fort Charlotte in March 1781 (1781) **150**
    Present Road leading Through the Garrison [and] Proposed Road to the Ferry (1785) **144**
    Plan of the Inclosed Battery or Redoubt near Leith, built for the protection of the Harbour in the Year 1780 (1785) **146**
    Sketch of the Ground around Fort George (1785) **169**
    Plan of the Bay called Brassa Sound (1786) **151**
French Revolutionary and Napoleonic Wars (1792–1815) 157, 163–64
Frew, Fords of 100
    William Edgar, *Description of the River Forth above Stirling* (1746) **102**
Fuller, John 185

Garelochhead 217
Geddie, John
    S. Andre sive Andreapolis Scotiae Universitas Metropolitana (*c.*1580) **25**
Geminus, Thomas 33
    The Englishe Victore agaynste the Schottes by Mvskelbroghe (1547) 22, **33**
Generalstab des Heeres *see* German Army / Generalstab des Heeres
Geographical Section General Staff *see* War Office / Geographical Section General Staff
George I, King 56, 57, 80
George II, King 80
George III, King (Topographical Collections) 53
German Army / Generalstab des Heeres 192–93, 197, 199, 201
    Oblique air photo – Blick von Queensferry über die Forthbrücke... (1940) **197**
    Oblique air photo – Borrowstouness (Bo'ness) (1940) **197**
    [Landing Beaches], Quarter-Inch to the mile with overprint (1940) **198–99**
    Bodenverhältnisse im Hinterland de Ostküste zwischen Humbermündung und Firth of Tay (Nördlicher Teil) (1940) **200**
    Schottland 1:50,000, Blatt Nr 64 – Dundee und St Andrews (1941) **201**
Glasgow 3, 131, 163, 164, 192, 193, 195
    see also Hillington
Glen Feshie
    Joseph Avery, [George Wade], *A Plan of the Country where the New Intended Road is to be made from the Barack at Ruthven in Badenoth to Invercall in Brae Marr...* (1735) **81**
Glen Shee
    George Morrison, *Survey of the Different Parts of the Road joyned betwixt Blair Gowrie and Brae Mar...* (1750) **138–39**, **152–53**
Glencoe 95
Glencorse 164
Glenelg *see* Bernera
Glengarry 56, 57
Glenshiel, battle of (1719) 57, 70, 73
    John Henri Bastide, *A Plan of the Field of Battle that was fought on ye 10th of Iune 1719, at the Pass of Glenshiels in Kintail...* (1719) **72–73**
global positioning systems (GPS) 184
Gordon, Harry 130, 135

Gordon, James
    Edinodunensis Tabulam... (1646) **29**
Gordon, Lewis (Marquis of Huntly) 39
'Gough' map 3
Graham, John (Viscount Dundee) 50
Graham, James (Marquis of Montrose) 38, 39–41, 47
Gravatt, William
    Sketch of the Harbour, Town & Battery of Arbroath (*c.*1795) **155**
Great Glen 50, 63, 84, 109, 116, 131, 196
    Joseph Avery, *This Plan Containing Lochness, Lochoyoch, Lochlochey, & all the Rivers and Strips of Water...* (1727) **90–91**
Greenock 155, 193
gunpowder magazines 70, 71, 86, 89, 93, 124, 126, 130, 131, 132, 142, 145, 147, 155
    David Watson, *Plan & Section of the Powder Magazine as it is at Present, Containing 684 Barrills of Powder* (1747) **132**
    David Watson, *Plan & Section of the Powder Magazine with the Alterations Propos'd* (1747) **132**
    William Skinner, *General Plan, Sections and Elevations of the Powder Magazine and Storehouses, Built in Edinburgh Castle Ano: 1753 & 1754* (1754) **133**
Guthrie, John (Bishop of Moray) 39

Hackness Martello *see* Longhope Sound
Hamilton, James (Regent Arran) 25
Hamilton, Sir James (of Finnart) 42
Hardyng, John 1–4, 211, 240
    Map of southern Scotland (*c.*1470s) **2–3**, 222
Hawley, Henry 97, 99, 104, 107
Henry II, King 24
Henry IV, King 3
Henry V, King 1
Henry VIII, King 24, 30, 52
Hertford, Earl of *see* Seymour, Edward (Earl of Hertford)
hillforts 7, 11
Hillington 186
    War Office / Geographical Section General Staff, Great Britain, 1:25,000 GSGS.3906, *Sheet 29/68 S.W.* (1941) **186**
    Ordnance Survey, Great Britain, 1:25,000 Provisional edition, *Sheet NS56* (1958) **187**
Hoddle, Robert
    Upper Plan of the Tower Built at Long Hope Sound (1815) **158**
    Section through... the Tower Built at Long Hope Sound (1815) **159**
Holinshed, Raphael
    Chronicles of England, Scotland and Ireland (1577) **28**
Hoy, Orkney *see* Longhope Sound
Hydrographic Office 16, 156–157, 167, 172–73, 193
    British Isles, English Channel and North Sea. Wreck Chart in 7 sheets. Sheet II. D20 (1919) **180**
    Waters surrounding the British Islands. Mined areas and safe channels (1919) **181**
    Gareloch. Admiralty Chart 2000 (1922) **204**
    Port Edgar to Carron. Admiralty Chart 114c (1925) **175**
    River Tay. Admiralty Chart 1481 (revised 1941) **189**
    Gareloch. Admiralty Chart 2000 (1943) **205**
    Admiralty Chart F.86: Approaches to Rosyth Dockyard (1944) **174–75**
    Gareloch. Admiralty Chart 2000 (2015) **216**

Inchgarvie 149, 154, 177
    Pawlett William Colebrooke, *Plan of the River or Firth of Forth* (1782) 148–49
Inchmickery 177
    War Office / Geographical Section General Staff, *Plan of Special Survey Inchmickery, Firth of Forth* (1918) **176–77**
International Civil Aviation Organisation (ICAO) 203
International Map of the World 210
Inverlochy *see* Fort William
Inveresk 31, 34
Inversnaid 67, 68, 126
    John Dumaresq, John Henri Bastide, *The Roads between Innersnait Ruthvan of Badenock Kiliwhiman and Fort William...* (1718) **68–69, 78–79**
    Andrews Jelfe, *Killewhiman, Inversnait, Ruthven of Badenoch, Bernera* (1719) **66**
    Andrews Jelfe, *Innersnait* (1719) **67**
    John Dumaresq, John Henri Bastide, *A Draught of Innersnait, in the Highlands of North Brittain...* (1719) **68**
Inverness 7, 15, 18, 38, 43, 56, 70, 74, 84–87, 91, 92, 96, 97, 98, 99, 100, 107, 116, 124, 126, 130–31
    see also Fort George (Inverness), 'Oliver's Fort' (Inverness)
    Lewis Petit, *Inverness in North Brittain* (c.1716) **74**
    John Henri Bastide, *A General Survey of Inverness, & the Country adjacent to the Foot of Loch-Ness; West Prospect of Inverness* (c.1725) **85**

Jacobite risings
    1689 Rising 47, 50, 129
    1708 Rising 56, 60–61
    1715 Rising 15, 56–57, 64, 65, 82–83, 95
    1719 Rising 57, 70, 72–73
    1745 Rising 18, 56, 63, 86, 89, 98–113
Jacobites/Jacobitism *see* Jacobite risings
jails *see* prisoners of war
James V, King 23
James VI, King 28, 36
James VII, King 6, 50
James VIII, King 56
Jelfe, Andrews 67, 70, 76, 84
    Killewhiman, Inversnait, Ruthven of Badenoch, Bernera (1719) **66**
    Innersnait (1719) **67**
    No. 1, Edinburgh Castle (1719) **76**
    No. 4, Edinburgh Castle (1719) **75**
Johnson, Robert 56
Johnson, Rowland 28
    Siege of Edinburgh Castle (1577) **28**
Jones, Jasper Leigh
    A Plan of ye Battle of Colloden... (1746) **110–11**
Jones, John Paul 131, 146

Keith, George (Earl Marischal) 70
Keppel, William (Earl of Albemarle) 80, 124, 126
Killiwhimen *see* Fort Augustus
Killiecrankie, battle of (1689) 47, 50
    Clement Lempriere, *A Description of the Highlands of Scotland...* (1731) **50**
Kincardine Castle
    Timothy Pont, Detail of Kincardine Castle, Strathearn, Pont 22 (c.1583–1614) **40**

John Adair, *The Mapp of Straithern, Stormount, and Cars of Gourie, with the Rivers Tay and Iern* (c.1720) **41**
Kinlochleven 130, 137
Kinloss 209
Kirkwood, David 163
Kirkwood, Robert
    Plan of the City of Edinburgh and its environs (1817) **165**

landing beaches 171, 210
    War Office. *Map to accompany the land defence of the Scottish zone* (1907) **170–71, 182–83**
    War Office. *Tay Defences, revised 1909* (1909) **171**
    German Army. *[Landing Beaches], Quarter-Inch to the mile with overprint* (1940) **198–99**
Lang Siege (Edinburgh Castle) 28
    Siege of Edinburgh Castle (1577) **28**
    James Gordon, *Edinodunensis Tabulam...* (1647) **29**
Laye, John
    A Plan of the Town and Castle of Sterling (1725) **8–9**
Lee, Richard 27, 28, 30
    The plat of Lythe w' th'aproche of the Trenchs therevnto (1560) **26**
Leith 23, 24, 26, 27, 30, 35, 38, 42, 43, 44, 131, 145, 146, 147, 149, 154, 157
    Richard Lee?, *The plat of Lythe w' th'aproche of the Trenchs therevnto* (1560) **26**
    Greenvile Collins, *Map of Leith from the North* (1693) **43**
    Alexander Wood, *... this plan of the town of Leith from an actual survey* (1777) **43**
    Andrew Frazer, *Plan of the Inclosed Battery or Redoubt near Leith, built for the protection of the Harbour in the Year 1780* (1785) **146**
    John Ainslie, *Old and New Town of Edinburgh and Leith with the proposed docks* (1804) **147**
Lemprière, Clement 70, 97
    A Description of the Highlands of Scotland... (1731) **50, 77**
Lennox, Charles (3rd Duke of Richmond) 149
Leuchars 190, 193
lighthouses 173, 179, 188, 195, 209, 213, 215
Limekilnburn
    Ordnance Survey, *Air Photo Mosaic 26/75 S.W.* (1946) **206**
    Ordnance Survey, *Air Photo Mosaic 26/75 S.W.* (1950) **207**
Linlithgow 99, 100
Linnhe, Loch 19, 51, 63
lithography 164
Littlejohn, Dr Henry 165
Loch Ness 68, 84, 91, 92
Loch Oich 98
Loch Tay 95
Lochaber 51, 95, 116
Loewenorn, Paul de 156
Longhope Sound 157, 159
    Philip Skene, *Plan of Long Hope Sound* (1815) **158**
    Philip Skene / Robert Hoddle, *Upper Plan of the Tower built at Long Hope Sound* (1815) **158**
    Philip Skene / Robert Hoddle, *Section through... the Tower built at Long Hope Sound* (1815) **159**
Lossiemouth 209
Louis XIV, King 54, 56
Louis XV, King 98, 113

Luce Bay 209
Luftwaffe 193, 195, 197, 202
Lyndsay, Alexander 23–24

MacDonald, Clan 95
MacDougall, Clan 19
Mackenzie, Kenneth Mòr (Earl of Seaforth) 44
Mackenzie, Murdoch 109
MacLean, Clan 19, 44
Macleod, Malcolm (Director-General of Ordnance Survey, 1935–43) 185
Manson, John 118
    Map of part of Scotland, showing clans that rebelled in 1715 (1731) 82–83
map colour/colouring 7, 54, 70, 71, 73, 74, 85, 119, 123,124, 126, 132, 141, 171, 185, 195, 199, 203, 215
map projections 7, 185, 210, 213
Marcell, Lewis 124
    Plans and profiles of buildings at 'Oliver's Fort', Inverness (1746) **125**
martello towers 12, 157–59
Mary II, Queen 50
Mary, Queen of Scots 24, 28
Melloni, Antonio 30
Military Revolution 30
Military Survey of Scotland 116, 118–19, 123–24, 145, 168
    William Roy, Detail from Dunkeld environs (1747– 52) **120–21**
    William Roy, Details from Stirling environs (1747– 55) **122**
Millan, J.
    A Plan of the Victory of Falkirk Muir Fought the Afternoon of January 16 1746 (1746) **106**
mines 173, 180
    Waters surrounding the British Islands. Mined areas and safe channels (1919 **181**
Minigaig Pass 68
Monadhliath Mountains 98
Monck, George (1st Duke of Albemarle) 38, 43
Mons Meg (gun)
    Munsmeg, a Gun so called at Edinburgh Castle (1734) **6**
Montrose, Marquis of *see* Graham, James (Marquis of Montrose)
Moore, Thomas
    No. 1, Edinburgh Castle (1719) **76**
    No. 4, Edinburgh Castle (1719) **75**
Moray Firth
    British Army of the Rhine, *Ascent Topographic Low Flying Chart: Europe 1:500.000. Sheet 7, United Kingdom, North* (1967) **208**
Morrison, George 135
    Survey of Part of the Road from Sterling to Fort William... (1749) **134–35**
    Survey of the different Parts of the Road joyned betwixt Blair Gowrie and Braemar... (1750) **138–39, 152–53**
Mull, Sound of 19
Munro, Sir George 39
Murray, Alexander, of Stanhope 109
Murray, William (Marquis of Tullibardine) 70, 73

Napoleonic Wars *see* French Revolutionary and Napoleonic Wars (1792–1815)
naval bases 172, 180, 203–4, 211, 213
news/newspaper maps 33, 99, 103, 192
Nicolay, Nicholas de 15, 23–24, 30
Vraye & exacte description Hydrographique des costes maritimes d'Escosse & des Isles Orchades... (1583) 27
Norham, Treaty of (1551) 36
North Sea 16, 210
    Hydrographic Office, *British Isles, English Channel and North Sea. Wreck Chart in 7 sheets. Sheet II. D20* (1919) **180**
    Hydrographic Office, *Waters Surrounding the British Islands: Mined Areas and Safe Channels* (1919) **181**
    Albert Close, *The naval war in the North Sea...* (1922) **16–17**
nuclear war 14, 211
    see also Faslane, Scottish Campaign for Nuclear Disarmament

O'Brien, John 63
'Oliver's Fort' (Inverness) 124, 131
    Lewis Marcell, Plans and profiles of buildings at 'Oliver's Fort', Inverness (1746) **125**
    William Skinner, Charles Tarrant, *Plan for building A Fort at Inverness on the Vestige of an old Fort demolished 1746* (1747) **124**
'Operation Review' 203
'Operation Sealion' 192
Ordnance Survey vii, 7, 11, 164, 168, 184, 185, 190, 192, 203, 207, 211, 217
    Large-Scale Town Plan, *Paisley, Sheet XII.3.11* (surveyed 1858, published 1863) **167**
    Large-scale town plan, *Edinburgh, Sheet 32* (revised 1877) **165**
    25 Inch to the mile, *Renfrew Sheet XII.3* (surveyed 1858, published 1864) **166**
    Six-Inch to the Mile, Roxburghshire, Sheet XXVIII (surveyed 1859, published 1863) **11**
    Six-Inch to the mile, *Ross and Cromarty, Sheet XC* (revised 1904, published 1907) **169**
    Six-Inch to the mile, *Fifeshire XLIII.NW* (revised 1924–5, published 1928) **175**
    Air Photo Mosaic 26/75 S.W. (1946) **206**
    Air Photo Mosaic 26/75 S.W. (1950) **207**
    OS Open Map - Local, detail of Faslane Port (2018) **217**
    One-Inch to the mile, Popular edition, *Sheet 68, Firth of Forth* (revised 1923–6, published 1928) **190**
    One-Inch to the mile, Popular edition with National Grid, *Sheet 68, Firth of Forth* (1924 with later revisions 1945) **191**
    Quarter-Inch to the mile. *Sheet 3 – The Forth & Tay* (1939) **21, 194**

Paisley Barracks 164, 166–67
    Ordnance Survey, 25-inch to the mile, *Renfrew, Sheet XII.3* (surveyed 1858, published 1864) **166**
    Ordnance Survey, Large-Scale Town Plan, *Paisley, Sheet XII.3.11* (surveyed 1858, published 1864) **167**
Paris, Matthew 3
Parkhall, Clydebank *see* Clydebank
Patten, William
    Battle of Pinkie (1548) **34**
Percy, Henry (Earl of Northumberland) 1
Perth 2, 7, 15, 38, 43, 57, 64, 65, 74, 96, 99, 100, 119, 131, 213
    prison 164
    Lewis Petit, *A Plan of Perth with the Retrenchment made about it by the Pretenders Engineers* (1716) **64**
    Lewis Petit, *Plan of Perth and Adjacent Places with a projection of a Cittadel* (1716) **64**

Petit, Lewis 15, 19, 64, 67, 71, 74,
  Plan of Castle Dwart in the Island of Moll... (1714) 19
  Plann of the Castle of Island Dounan; Profile of the Front of the Castle of Island Dounan (1714) 71
  A Plan of Perth with the Retrenchment made about it by the Pretenders Engineers (1716) 64
  Plan of Perth and Adjacent Places with a projection of a Cittadel (1716) 64
  Inverness in North Brittain (c.1716) 74
Petit, Thomas 15, 30, 31
  The platte of Milkcastle (1547) 32
Philip V, King 57
Philiphaugh, battle of (1645) 40
photo-lithography/zincography *see* lithography
Piershill Barracks
  Robert Kirkwood, *Plan of the City of Edinburgh and its environs* (1817) 165
  Ordnance Survey, Large-scale town plan, *Edinburgh, Sheet 32* (revised 1877) 165
Pinkie, battle of (1547) 27, 30–31
  Thomas Geminus alias Lambrechts, *The Englishe Victore agaynste the Schottes by Mvskelbroghe* (1547) 22, 33
  William Patten, *Battle of Pinkie* (1548) 34
  John Ramsay, Battle of Pinkie commemorative scroll (c.1548?) 34–35, 48–49
Pinkie Cleugh 24, 31
place names vii, 3, 156, 215, 217
Polaris missile system *see* Faslane
Pont, Timothy 12, 36, 38
  Timothy Pont, Detail of Duffus Castle and Spynie Palace, [Gordon] 23 (c.1583–1614) 39
  Timothy Pont, Detail of Kincardine Castle, Strathearn, Pont 22 (c.1583–1614) 40
  Timothy Pont / Joan Blaeu, *Lidalia vel Lidisdalia regio, Lidisdail* (1654) 36
Preston, Thomas 156
Prestonpans, battle of (1745)
  A plan of the battle of Preston Panns fought 21st Sept. 1745 (1745) 103
prisoners of war 7, 157, 163–64
  Charles Tarrant, *Plans and Sections of the Several Vaults . . . in Edinburgh Castle...* (1754) 163–64
Privy Council 56
projections *see* map projections
Protestant Truth Society 16
Ptolemy, Claudius 3, 11

quarries 53, 210
Queensferry 154, 195
  Pawlett William Colebrooke, *Plan of the River or Firth of Forth* (1782) 148–49
  Oblique air photo – Blick von Queensferry über die Forthbrücke... (1940) 197

radar 173, 206–7
Radical War 164, 166
railways 185, 188, 190, 192, 199, 217
Ramsay, John
  Battle of Pinkie commemorative scroll (c.1548?) 34–35, 48–49
Ramsden, Jesse 168
reconnaissance 1, 4, 36–38, 57, 74, 81, 83, 119, 168, 184, 192–193, 211, 215
  see also surveys and surveying

Redford Barracks 165
Richard II, King 3
Riccarton
  Soviet Army, General Staff, *Velikobritaniia, Shotlandiia. Edinburg (N-30-6)* (1983) 215
Ridgeway, William 31
riot control 164–65
roads, civilian 165, 166, 169, 175, 185, 199, 210, 213, 215, 217
roads, military 50, 57, 68, 70, 77, 81, 83, 84, 91, 92, 95–98, 99, 100, 107, 109, 118, 119, 129, 130–31, 135–39, 144
  George Wade, *Map showing the intended military roads joining up Stirling with Fort Augustus, etc.* (1725) 96
  Richard Cooper, *A map of His Majesty's Roads from Edinburgh to Inverness, Fort Augustus & Fort William* (c.1742) 97–98
  Andrew Rutherford, *An Exact Plan of His Majesty's Great Roads through the Highlands of Scotland* (1745) 100–101
  George Morrison, *Survey of Part of the Road from Sterling to Fort William...* (1749) 134–35
  John Archer, *A Survey of the Road made by Coll Rich's & Genl Guise's regts between Fort William & the head of King-loch Leven* (1750) 136–37
  George Morrison, *Survey of the different Parts of the Road joyned betwixt Blair Gowrie and Braemar...* (1750) 138–39, 152–53
  Andrew Frazer, *Present Road leading Through the Garrison [and] Proposed Road to the Ferry* (1787) 144
Robert Gordon's College 13
Rogers, John 30
Romans 7, 11, 70, 77
Romer, John Lambertus 61, 84, 85, 86, 88, 89, 93
  A Plan of Fort George, & part of the Town of Inverness, with proper Sections relating to the Fort (1732) 86–87
  A Plan of Edinburgh Castle (1737) 61
  A plan of the intended Fortress with the Situation of Killiwhymen (1729) 88
  Fort William; Fort Augustus; Fort George, Inverness (c.1729–46) 114–15
  The Plan of Fort Augustus, in the Highlands of Scotland... (c.1742) 89
Romer, Wolfgang William 84
Rosetti, Giovanni di 30
Rosneath 203–4, 211
Rosyth 172, 175, 196, 213
  Ordnance Survey, Six-Inch to the mile, *Fifeshire XLIII.NW* (revised 1924–5, published 1928) 175
  Hydrographic Office, *Port Edgar to Carron. Admiralty Chart 114c* (1925) 175
  Hydrographic Office, *Admiralty Chart F.86: Approaches to Rosyth Dockyard* (1944) 174–75
Rotz, Jean 30
Rough Wooing, War of the (1544–50) 1, 20, 24, 30–36
Roy, William 7, 11, 20, 99, 116, 118–19, 123, 168
  Military Survey of Scotland (1747–55) 120–22
  Plan of the Environs of the Eildon Hills on the South Bank of the Tweed... (1793) 10
Royal Engineers 158, 171, 173, 204
Royal Military Academy, Woolwich 52, 118
Rudyard, Henry
  Plan of Part of Edinburgh Castle Showing the Proposed Situation for a Pipe & Cistern (1794) 160–61
Russian Army / Russian military maps *see* Soviet Army, General Staff

INDEX

Rutherford, Andrew
 An Exact Plan of His Majesty's Great Roads through the Highlands of Scotland (1745) 100–101
Ruthven 57, 67, 81, 98, 100, 107, 126
 John Dumaresq, John Henri Bastide, *The Roads between Innersnait Ruthvan of Badenock Kiliwhiman and Fort William...* (1718) 68–69, 78–79
 Andrews Jelfe, *Killewhiman, Inversnait, Ruthven of Badenoch, Bernera* (1719) 66
rutter ('routier') 23–24

St Andrews 23–24
 John Geddie?, *S. Andre sive Andreapolis Scotiae Universitas Metropolitana* (c.1580) 25
Sandby, Paul 19, 119, 126
 Plan of the Castle of Dunbarton (c.1747) 127
Sark, River 35, 36, 130
satellite imagery 184, 207, 211, 217
Schultz, Henry
 Plan of the Battle of Collodden (1746) 109
Scotland, maps of
 John Hardyng, *Map of Scotland* (1457) 2–3, 222
 Nicolas de Nicolay, *Vraye & exacte description Hydrographique des costes maritimes d'Escosse & des Isles Orchades...* (1583) 27
 John Elphinstone, *A New Map of North Britain...* (1746) 80, 104
 John Elphinstone, David Watson, *A New & Correct Mercator's Map of North Britain* (1746) 108
 Evacuation Map of Scotland, *The Scotsman* (1939) 196
 Soviet Army, General Staff, *Velikobritaniia, Shotlandiia. Aberdin O-30* (1986) 212
Scott, Sir Walter 6
Scottish Campaign for Nuclear Disarmament
 Scotland: no place for nuclear weapons (2014) 217
Second World War 184, 186, 189, 192–93, 195, 203
 see also German Army / Generalstab des Heeres
 War Office / Geographical Section General Staff, *General plan of Barry Links* (1939) 188
 Hydrographic Office, *River Tay. Admiralty Chart 1481* (revised 1941) 189
 War Office / Geographical Section General Staff, *Great Britain, 1:25,000 GSGS.3906, Sheet 29/68 S.W.* (1941) 186
 Ordnance Survey, *Great Britain, 1:25,000 Provisional edition, Sheet NS56* (1958) 187
Seymour, Edward (Earl of Hertford / Lord Protector Somerset) 23–24, 27, 30–31, 34
Sheriffmuir, battle of (1715) 57, 65
 Plan of the Battle of Sheriffmuir (1715) 65
Shetland
 Dépôt Général de la Marine, *Carte des Îles Shetland* (1806) 156
signal stations 157, 159, 171
Skene, Philip
 Plan of Long Hope Sound (1815) 158
 Upper Plan of the Tower built at Long Hope Sound (1815) 158
 Section through... the Tower built at Long Hope Sound (1815) 159
Skinner, Monier 126
Skinner, William 15, 118, 124, 126, 131, 141, 161
 Plan for building A Fort at Inverness on the Vestige of an old Fort demolished 1746 (1747) 124
 Plan of the point of land at Arderseer with the design'd fort as trac'd thereon (1748) 140–41
 Plans Section & Elevation of the Barrack at Corgarff (c.1750) 129
 A Plan of Fort George, North Britain, shewing how far executed (1753) 142
 Fort George, Ardersier – sections (1753) 143
 General Plan, Sections and Elevations of the Powder Magazine and Storehouses, Built in Edinburgh Castle Ano: 1753 & 1754 (1754) 133
Skye 53, 71, 109
Slezer, John 24, 38, 47, 51, 54, 56, 58–59
 Plan of Dunotter (1675) 46
 Innerlochie or Obrian Fort, in Lochabor; Desine for Innerlochie (1689) 51
 A coloured bird's-eye view of Edinburgh Castle, showing the projected outworks (c.1690) 54–55
 Plans and Prospects of Stirling Castle (c.1696) 58–59
Society of Antiquaries of Scotland 6
South Walls, Orkney *see* Longhope Sound
Soviet Army, General Staff 203, 210–15
 Velikobritaniia, Shotlandiia. Aberdin O-30 (1986) viii, 212
 Velikobritaniia, Shotlandiia. Kerkkoldi O-30-V (1986) 213
 Aberdin (O-30-104) (1981) 214
 Velikobritaniia, Shotlandiia. Edinburg (N-30-6) (1983) 215
spravka (geographical description) 210
Spynie Castle 39
 Timothy Pont, *Detail of Duffus Castle and Spynie Palace, [Gordon] 23* (c. 1583–1614) 39
Steòrnabhagh *see* Stornoway/Steòrnabhagh
Stewart, James (Regent Moray) 25
Stirling 3, 7, 30, 38, 50, 56, 58–59, 60, 65, 70, 92, 96, 98–102, 104, 123, 126, 130–31, 157
 John Slezer, *Plans and Prospects of Stirling Castle* (c.1696) 58–59
 Theodore Dury, Talbot Edwards, *A Plan of Sterling Castle...; A profile of Elphinstons Tower and French Spur at Sterling Castle* (c.1708) 60
 John Laye, *A Plan of the Town and Castle of Sterling* (1725) 8–9
Stornoway/Steòrnabhagh 38, 70
 The ground platt of the fortification at Stornoway upon Lewis Island (1653) 44–45
Strath Braan
 William Roy, *Military Survey of Scotland* (1747–1752) 120–21
Strath Spey 68
 Joseph Avery, [George Wade], *A Plan of the Country where the New Intended Road is to be made from the Barack at Ruthven in Badenoth to Invercall in Brae Marr...* (1735) 81
Strathyre 135
 George Morrison, *Survey of Part of the Road from Sterling to Fort William...* (1749) 134–35
Strauss, Rear-Admiral Richard 180
Strozzi, Leon 24–25
Strozzi, Pietro 30, 35
Stuart, Prince Charles Edward (Bonnie Prince Charlie/The Young Pretender) 80, 84, 96, 99, 107, 113, 116,
Stuart, Francis (Earl of Moray) 102
Stuart, Henry (Lord Darnley) 25
submarines 12, 14, 16, 154, 172–73, 184–85, 203, 211, 216
surveys and surveying 4, 7, 15, 19, 24, 31, 36, 38, 47, 52, 56, 59–60, 63, 68, 81, 84, 85, 88, 91, 95, 97, 107, 118–119, 124, 126, 130, 135, 141, 156, 158, 168, 175, 203
 see also reconnaissance

231

Tarrant, Charles 18, 119, 126
- *A Plan of Sterling Castle...; A profile of Elphinstons Tower and French Spur at Sterling Castle* (c.1708) 60
- *Plan of Fort George at Inverness, Shewing it's present Condition* (1750) 18
- *Plan for building A Fort at Inverness on the Vestige of an old Fort demolished 1746* (1747) **124**
- *Plan of the Point of Land at Arderseer with the Design'd Fort as Trac'd Thereon* (1748) **140–41**
- *Plans Section & Elevation of the Barrack at Corgarff* (c.1750) **129**
- *A Plan of Fort George, North Britain, Shewing how Far Executed* (1753) **142**
- *Fort George, Ardersier – sections* (1753) **143**
- *Plans and Sections of the Several Vaults and Floors... in the Edinburgh Castle 1754* (1754) **162–63**

Tay, Firth of
- War Office / Geographical Section General Staff, *Map to accompany the land defence of the Scottish zone* (1907) **170–71, 182–83**
- War Office / Geographical Section General Staff, *Tay Defences, revised 1909* (1909) **171**
- [Landing Beaches], Quarter-Inch to the mile with overprint (1940) **198–99**
- Hydrographic Office, *River Tay. Admiralty Chart 1481* (revised 1941) **189**

Tentsmoor 189
- Hydrographic Office, *River Tay. Admiralty Chart 1481* (revised 1941) **189**

Three Kingdoms, Wars of the (1639–53) 14, 38, 41
topography 7, 36–38, 57, 74, 102, 104, 107, 118, 185
town plans 7, 15, 74
- Lewis Petit, *Inverness in North Brittain* (1716) **74**
- John Laye, *A Plan of the Town and Castle of Sterling* (1725) **8–9**
- Alexander Wood, *To the Magistrates, the Commissioners of Police and the Four Incorporations, this Plan of the Town of Leith from an Actual Survey* (1777) **43**
- Ordnance Survey, Large-Scale Town Plan, *Paisley, Sheet XII.3.11* (surveyed 1858, published 1864) **167**
- Ordnance Survey, Large-scale town plan, *Edinburgh, Sheet 32* (revised 1877) **165**

trace italienne fortification 7, 27, 30, 35
training camps 12, 14, 131, 168, 185, 188, 203
Tullibardine Castle
- Timothy Pont, Detail of Kincardine Castle, Strathearn, Pont 22 (c.1583–1614) **40**
- John Adair, *The Mapp of Straithern, Stormount, and Cars of Gourie, with the Rivers Tay and Iern* (c.1720) **41**

Turnhouse 190, 195

Ubaldini, Migiliorino 30, 35
Umfraville, Robert (Earl of Kyme) 1
Union, Act of (1707) 12, 20, 47, 56

Vauban, Sébastian le Prestre de 55

Wade, George 57, 68, 80–84, 88, 90, 91–99, 130–31
- *Map showing the intended military roads joining up Stirling with Fort Augustus, etc.* (1725) **96**
- *A plan of the intended Fortress with the Situation of Killiwhymen* (1729) **88**
- *A Plan of the Country where the New Intended Road is to be made from the Barack at Ruthven in Badenoth to Invercall in Brae Marr...* (1735) **81**
- *A Plan of Fort William in the Shire of Inverness* (1736) **62**

War Office / Geographical Section General Staff 15, 126, 168, 171, 185, 186
- *Map to accompany the land defence of the Scottish zone* (1907) **170–71, 182–83**
- *Tay Defences, revised 1909* (1909) **171**
- *Plan of Special Survey Inchmickery, Firth of Forth* (1918) **176–77**
- *Clyde, Cloch Point* (1918) **178–79**
- *General plan of Barry Links* (1939) **188**
- *Great Britain, 1:25,000 GSGS.3906, Sheet 29/68 S.W.* (1941) **186**

water supply 7, 141, 157, 159, 161, 210

Watson, David 20, 109, 118–19, 124, 126, 132
- *A New & Correct Mercator's Map of North Britain* (1746) **108**
- *Plan & Section of the Powder Magazine as it is at present, containing 684 Barrills of Powder* (1747) **132**
- *Plan & Section of the Powder Magazine with the alterations propos'd* (1747) **132**

Wharton, Thomas (Warden of the West Marches) 31
Wightman, Joseph 70, 73
William III (of Orange), King 47, 50, 57, 84
Woden Law (hillfort)
- Ordnance Survey, Six-Inch to the Mile, Roxburghshire, Sheet XXVIII (surveyed 1859, published 1863) **11**

wrecks 180
- Hydrographic Office, *British Isles, English Channel and North Sea. Wreck Chart in 7. sheets. Sheet II. D20* (1919) **180**

Wood, Alexander
- *... this plan of the town of Leith from an actual survey* (1777) **43**

Yorke, Joseph 107
Young Pretender *see* Stuart, Prince Charles Edward (Bonnie Prince Charlie / The Young Pretender)